# PHASE DIAGRAMS IN METALLURGY

## Their Development and Application

Metallurgy and Metallurgical Engineering Series

ROBERT F. MEHL, *Consulting Editor*
MICHAEL B. BEVER, *Associate Consulting Editor*

---

BARRETT · Structure of Metals
BIRCHENALL · Physical Metallurgy
BRIDGMAN · Studies in Large Plastic Flow and Fracture
BRIGGS · The Metallurgy of Steel Castings
BUTTS · Metallurgical Problems
DARKEN AND GURRY · Physical Chemistry of Metals
DIETER · Mechanical Metallurgy
HANSEN · Constitution of Binary Alloys
KEHL · The Principles of Metallographic Laboratory Practice
RHINES · Phase Diagrams in Metallurgy
SEITZ · The Physics of Metals
SMITH · Properties of Metals at Elevated Temperatures
WILLIAMS AND HOMERBERG · Principles of Metallography

# Phase Diagrams in Metallurgy

## THEIR DEVELOPMENT AND APPLICATION

**FREDERICK N. RHINES, Ph.D.**

*Aluminum Company of America*
*Professor of Light Metals and Member of the*
*Staff of the Metals Research Laboratory*
*Carnegie Institute of Technology*

McGRAW-HILL BOOK COMPANY, INC.

New York     Toronto     London

1956

PHASE DIAGRAMS IN METALLURGY

*Library of Congress Catalog Card Number 56–8272*

VI

52070

THE MAPLE PRESS COMPANY, YORK, PA.

# PREFACE

Phase diagrams mean more to the metallurgist than mere graphical records of the physical states of matter. They provide a medium of expression and thought that simplifies and makes intelligible the otherwise bewildering pattern of change that takes place as elemental substances are mixed one with another and are heated or cooled, compressed or expanded. They illumine relationships that assist us in our endeavor to exercise control over the behavior of matter. These are no accidental by-products of a system devised primarily for the recording of physical data. They exist because the basic features of construction of all phase diagrams are dictated by a single natural law, the phase rule of J. Willard Gibbs, which relates the physical state of a mixture with the number of substances of which it is composed and with the environmental conditions imposed upon it.

Gibbs first announced the phase rule in 1876, but it was not until the opening of the present century that its significance was appreciated by any substantial fraction of the scientific world. To Bakhuis Roozeboom is due much of the credit for interpreting the phase rule to the scientific public and for demonstrating its usefulness in many of the branches of endeavor in which it is now applied. His celebrated monograph "Die Heterogenen Gleichgewichte vom Standpunkte der Phasenlehre," the first portion of which was published in 1900, is a comprehensive compilation and interpretation of all that was known at the time concerning the constitution of all kinds of chemical systems, metallurgical and otherwise. The completion of this gigantic work, after Roozeboom's death, was due mainly to the efforts of F. A. H. Schreinemakers, E. H. Buchner, and A. H. W. Aten.

Since that time a number of textbooks and monographs on the phase rule have appeared. Three of the most recent of these have been written by metallurgists, for the specific use of metallurgists, namely: G. Masing's "Ternäre Systeme," first published in the German language in 1932 and subsequently (1944) translated into English by B. A. Rogers; J. S. Marsh's "Principles of Phase Diagrams," issued in 1935 as a contribution to the Alloys of Iron Monograph Series; and R. Vogel's "Die Heterogenen Gleichgewichte," a modernized and greatly improved version of Roozeboom's book, published in 1937 as the second volume of the "Handbuch der Metallphysik."

In addition, there have been several comprehensive collections of phase diagrams of the alloy systems. One of the first of these, now obsolete, appeared in the second volume of the "International Critical Tables." In 1936 M. Hansen published the "Aufbau der Zweistofflegierungen," a critical survey of all binary phase diagrams of metal systems that has become the standard reference book on this subject throughout the world. Lesser collections have appeared in several textbooks and handbooks, notably the "Metals Handbook" of the American Society for Metals, the most recent issue of which carries an abridged collection of diagrams that have been revised to conform with the most reliable information available in 1946. The only serious attempt to produce an exhaustive collection of ternary and higher order diagrams is that of E. Janeck, "Handbuch Aller Legierungen," published in 1937. Most of these works include extensive reference to the research literature. A conveniently arranged and nearly exhaustive bibliography on the constitution of alloy systems has been compiled by J. L. Haughton and has been published by The Institute of Metals of Great Britain.

The present book is derived from a series of lectures presented over a period of more than two decades to undergraduate students in the Department of Metallurgical Engineering at the Carnegie Institute of Technology. Some years ago the lecture notes were issued as a mimeographed booklet that has enjoyed modest circulation, having been translated and published in the Portuguese by João Mendes Franca of the Institute of Technologic Research, at São Paulo, Brazil. This reception has encouraged the author to recast the notes in book form.

If this book can be said to have a special mission, a justification for its addition to the literature of phase diagrams, it is to present the subject in a manner that can be grasped by undergraduate engineering students whose primary interest is in the application of phase diagrams to metallurgical problems. To this end, the treatment of thermodynamic principles has been reduced to a minimum and those types of equilibria which are of special importance to chemists but are rarely encountered in metal systems have been deemphasized in favor of those of more immediate interest to metallurgists.

A particular effort has been made to anticipate and to answer those questions which have been observed to arise in the minds of students when first confronted with the various concepts and conventions used in the representation and the interpretation of phase equilibria. This has led to the inclusion of rather frequent excursions into bordering fields of physical metallurgy. Special attention has been devoted also to the pacing of the presentation, which commences at a slow tempo that increases gradually until the later chapters proceed at a rate that presupposes a mastery of the basic principles.

The original lecture notes covered the content of Chaps. 2 to 18, inclusive. An introductory chapter has been inserted for the assistance of those who may have had no previous contact with the subject. Additional chapters on the representation of multicomponent systems and of pressure-temperature diagrams have been included in response to a growing need for means to deal with complex alloy systems and especially those involving gaseous elements. Finally, and as a compromise with the older pedagogy that approached the subject from the viewpoint of the *determination* of phase diagrams rather than their *applications*, a brief chapter on research methods has been included.

Experience indicates that a working knowledge of phase diagrams cannot be imparted by the written or spoken word alone. Just as any mathematical subject must be practiced to be mastered, so the subject of phase diagrams requires practice on the part of the student if he is to obtain anything more than a superficial glimpse into the field. Some problems are given at the ends of several of the chapters; these are intended as suggestions only. The most informative problems are often derived from questions that arise in the mind of the reader. If he will take the time to answer these for himself by applying the principles of phase-diagram construction and interpretation that are presented in the book, he will find himself repaid many times over in progress toward a mastery of the subject.

The manipulation of phase diagrams calls into play some faculties that are rarely employed in other branches of scientific endeavor. Chief among these is space perception, as encountered in three-dimensional diagrams. Those who have had little previous reason to develop this faculty will usually be assisted by mechanical aids to visualization, such as actual three-dimensional models of the diagrams. Some suggestions with regard to ways in which such models may be constructed with a minimum of expense and effort are given in Chap. 18.

In closing, the author wishes to express his gratitude to the people who have assisted with the preparation of illustrations for this book: to Mrs. Wilma Urquhart, Mrs. Miles Price, and Mr. Lloyd Hughes, who prepared most of the photomicrographs; to Dr. Malcolm F. Hawkes, through whose kind permission the photomicrographs of Figs. 5–5, 5–7, 5–8, 5–11, 5–12, 5–13, and 5–15 are reproduced; to Professor C. Muhlenbruch, who supplied the photographs reproduced in Fig. 5–3; to Dr. Chester W. Spencer, who prepared the photomicrographs used in Figs. 9–3, 9–4, and 10–3; and to Felix Cooper for his drawings of three-dimensional subjects.

FREDERICK N. RHINES

# CONTENTS

# CHAPTER 1

# INTRODUCTION

When a metal melts it is said to undergo a *phase change;* the *solid phase* is transformed into the *liquid phase.* Other phase changes occur at the boiling point, where the *liquid phase* is transformed into the *gas phase;* at the sublimation point, where the *solid phase* is transformed into the *gas phase;* and at the temperature of allotropic transformation, where one kind of *solid phase* is changed into another kind of *solid phase.* The solid phases are always crystalline, and differences among solid phases are differences in composition, crystal structure, or crystal dimensions.

Under ordinary conditions, with the pressure constant, the phase changes in pure metals occur *isothermally;* that is, melting takes place at a single definite melting temperature, boiling at a fixed boiling point, and so on. Certain alloys likewise undergo *isothermal phase changes,* but it is more frequently found that the phase changes in alloys occur, instead, over ranges of temperature. Melting, for example, may begin at one temperature and not be completed until some higher temperature is reached, the alloy meanwhile existing in a mushy state composed of mixed liquid and solid. More complex phase changes, both isothermal and nonisothermal, are common. It is often found, for example, that more than two phases are involved in a single transformation. Thus, a molten alloy may freeze in such a way as to deposit a complex solid composed of several different solid phases.

Through the years an enormous body of information concerning the phase changes in alloys has been accumulated. Not only many thousands of commercial alloys but also a vastly greater number of alloys not in common use have been examined in this respect. Some systematic method of recording this information is required in order to condense it and to make available such individual items of data as may be needed in the daily handling of metals. The most successful method yet devised to accomplish this is the use of *phase diagrams,* also known as *constitutional diagrams* or *equilibrium diagrams.*

## The Phase Diagram

A typical phase diagram, presented in Fig. 1-1, indicates the phases present in all possible alloys of the two metals nickel (Ni) and copper (Cu), at all temperatures from 500 to 1500°C. Alloy composition is given

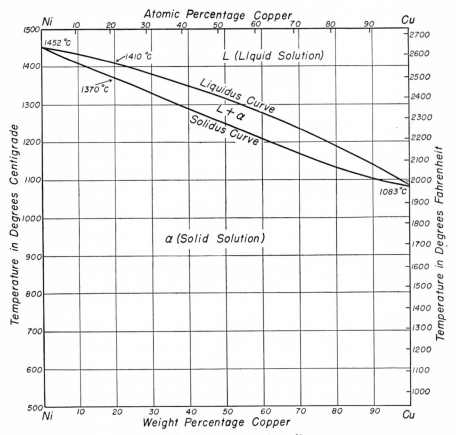

FIG. 1-1. The nickel-copper phase diagram.

by the horizontal scale, along the base of the diagram, where the percentage, by weight, of copper is read directly, the remainder being nickel. Temperature is read vertically, either from the centigrade scale on the left or from the Fahrenheit scale on the right. Two curves cross the diagram from the melting point of nickel at 1452°C to the melting point of copper at 1083°C. Of these, the upper curve, called the *liquidus*, denotes for each possible alloy composition the temperature at which freezing begins during cooling or, equivalently, at which melting is completed upon heating. The lower curve, called the *solidus*, indicates the temperatures at which melting begins upon heating or at which freezing is completed upon cooling. Above the liquidus every alloy is molten, and this region of the diagram is, accordingly, labeled "*L*" for liquid phase or liquid solution. Below the solidus all alloys are solid, and this region is labeled "*α*" because it is customary to use a Greek letter (see Appendix I) for the designation of a solid phase which is also a solid solution. At temperatures between the two curves the liquid and solid phases are present together, as is indicated by the designation "*L* + *α*."

Thus, the melting range of any desired alloy, say an alloy composed of 20% copper (balance, 80% nickel), may be found by tracing the vertical line originating at 20% copper on the base of the diagram to its intersections with the solidus and liquidus. In this way, it will be found that the alloy in question begins to melt at 1370°C and is completely molten at 1410°C.

## Temperature and Composition Scales

Custom favors the use of the centigrade scale of temperature in the construction of phase diagrams. The Fahrenheit scale may be used alternatively, however, as convenience dictates (see temperature-conversion table, Appendix IV). It is likewise most common to express alloy composition in "weight percentage," but for certain types of scientific work the "atomic percentage" scale (top edge of Fig. 1) may be preferred. The scale upon which the composition is expressed makes no fundamental difference in the form of the resulting phase diagram. If desired, composition may also be given in terms of the percentage by volume, but this usage is rare in the representation of metal systems.

The conversion from weight percentage (wt. %) to atomic percentage (at. %) or the reverse may be accomplished by the use of the following formulas:

$$\text{At. } \% \ X = \frac{(\text{wt. } \% \ X)/(\text{at. wt. } X)}{(\text{wt. } \% \ X)/(\text{at. wt. } X) + (\text{wt. } \% \ Y)/(\text{at. wt. } Y)} \times 100$$

$$\text{Wt. } \% \ X = \frac{(\text{at. } \% \ X)(\text{at. wt. } X)}{(\text{at. } \% \ X)(\text{at. wt. } X) + (\text{at. } \% \ Y)(\text{at. wt. } Y)} \times 100$$

where $X$ and $Y$ represent the two metals in the alloy. Tables for making these conversions (useful when a large number of such computations is to be made) and also formulas for ternary and higher order systems are given in Appendix III. A table of atomic weights (at. wt.) appears in Appendix II.

## Equilibrium

All properly constructed phase diagrams record the phase relationships only as they occur under conditions of *equilibrium*. This is necessary because phase changes as observed in practice tend to occur at different temperatures, depending upon the rate at which the metal is being heated or cooled. With rapid heating, any phase change, such as melting, occurs at a slightly higher temperature than with slow heating. Conversely, with rapid cooling the phase change occurs at a lower temperature than with slow cooling. Thus, transformations observed during heating are at higher temperature than the reverse transformations observed during cooling, except in the hypothetical case wherein the rates of heating and cooling are infinitely slow, whereupon the two observations of temperature would coincide at the *equilibrium transformation temperature*.

The equilibrium states that are represented upon phase diagrams are known as *heterogeneous equilibria*, because they refer to the coexistence of different states of matter. For two or more phases to attain mutual equilibrium, however, it is necessary that each be internally in a homogeneous state. In general, this means that each phase must be in the lowest energy state of which it is capable under the restrictions imposed by its environment. Thus, the chemical composition must be identical everywhere within the phase, the molecular and atomic species of which the phase is composed (if more than one) must be present in equilibrium proportions, and crystalline phases must be free of internal stresses.

An exception to the rule that only true equilibrium states are recorded on phase diagrams is found in the occasional representation of so-called *metastable equilibria*. In ordinary carbon steels, for example, there is found a solid phase, a carbide of iron ($Fe_3C$), that decomposes into graphite and iron under conditions that are favorable to the attainment of true equilibrium. The rate of decomposition of the iron carbide is very slow, however, under the most favorable conditions and is usually imperceptible under ordinary conditions. Because of its reluctance to decompose, this phase is said to be metastable, and it is represented on the usual (metastable) iron-carbon phase diagram. Evidently, metastability is a concept incapable of definition except by fiat, because there is no fundamental basis for saying that those substances that revert to the stable form at less than a certain rate are *metastable* while those that decompose more rapidly are *unstable*. The recognition of "metastable phase diagrams" is simply a practical artifice that has been found useful in certain instances, even though in violation of the basic assumptions of the phase rule.

## The Phase Rule

The construction of phase diagrams is greatly facilitated by certain rules which come from the science of thermodynamics. Foremost among these is Gibbs' phase rule.[1] This rule says that the maximum *number of phases P* which can coexist in a chemical system, or alloy, plus the *number of degrees of freedom F* is equal to the *sum of the components C* of the system plus 2.

$$P + F = C + 2$$

The *phases P* are the homogeneous parts of a system which, having definite bounding surfaces, are conceivably separable by mechanical means alone, for example, gas, liquid, and solid.[2]

*The degrees of freedom F* are those externally controllable conditions of temperature, pressure, and composition which are independently variable and which must be specified in order to define completely the state of the system at equilibrium.[3]

The *components C* are the smallest number of substances of independently variable composition making up the system. In alloy systems, it is usually sufficient to count the number of elements present; in the case of a mixture of stable compounds, such as salt and water, the number of components may be taken as two ($NaCl + H_2O$), unless carried to a degree of temperature and pressure where one or both of the compounds

[1] Formal derivations of the phase rule will be found in a large number of available texts, including J. S. Marsh, "Principles of Phase Diagrams," p. 24, McGraw-Hill Book Company, Inc., New York, 1935; A. C. D. Rivett, "The Phase Rule," chap. I, Oxford University Press, New York, 1923; A. Findlay, "The Phase Rule," p. 13, Longmans, Green & Co., Ltd., London, 1931, and pp. 17–19, Longmans, Green & Co., Inc., New York, 1951; L. Page, "Introduction to Theoretical Physics," p. 277, D. Van Nostrand Company, Inc., New York, 1928; John E. Ricci, "The Phase Rule and Heterogeneous Equilibrium," pp. 14–18, D. Van Nostrand Company, Inc., New York, 1951.

[2] In succeeding chapters the term *phase* will often be found used in a looser sense to designate the several states of matter in an alloy system, whether these are individually "homogeneous" or not. It should be understood that within its precise meaning the term *phase* can be applied only to states of matter at equilibrium and, hence, necessarily homogeneous.

[3] The assumption is here made that temperature, pressure, and concentration are the only externally controllable variables capable of influencing the phase relationships. Such variables as the electrostatic field, the magnetic field, the gravitational field, and surface-tension forces are considered to have no appreciable influence. If one of these variables should become important, in a specific instance, it would be necessary to include it among the externally controllable variables and to increase the constant in the phase rule from 2 to 3, that is, $P + F = C + 3$.

decomposes, when it becomes necessary to consider four components (Na, Cl, H, and O).

Through numerous examples in subsequent chapters the meaning of these somewhat formidable definitions will become clear. For the present, however, it will be sufficient to illustrate the application of the phase rule with a simple example. Suppose that it is desired to ascertain under what conditions a pure metal can exist with the gas, liquid, and solid phases all present in a state of equilibrium. There are, then, three phases. Since only one metal is involved the number of components is one. The phase rule is equated to find the number of degrees of freedom, thus:

$$P + F = C + 2$$
$$3 + 0 = 1 + 2$$

There are seen to be no degrees of freedom, which means that the coexistence of these three phases can occur only at one specific temperature and one specific pressure (the composition is, of course, fixed, only one metal being present). If one such set of specific conditions of temperature and pressure is now found by experiment, it will be unnecessary to look for another set, because the phase rule shows that only one can exist. Moreover, if it is desired to construct a phase diagram in which the coexistence of the three phases is represented, it becomes apparent at once that the coordinates of the diagram should be temperature and pressure and that the coexistence of the three phases must be indicated by a single point on this diagram (see point $O$ in Fig. 1-2).

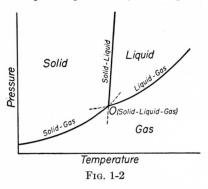

Fig. 1-2

## Other Rules

Although the phase rule tells what lines and fields should be represented upon a phase diagram, it does not usually define their shapes or the directions of the lines. Further guidance in the latter respect may be had from several additional rules of thermodynamic derivation.

The *theorem of Le Châtelier*, for example, says that *if a system in equilibrium is subjected to a constraint by which the equilibrium is altered, a reaction takes place which opposes the constraint, i.e., one by which its effect is partially annulled* (Ostwald). Thus, if an increase in the temperature of an alloy results in a phase change, that phase change will be one which proceeds with heat absorption, or if pressure applied to an alloy system brings about a phase change, this phase change must be one which is accompanied by a contraction in volume.

The usefulness of this rule can again be shown by reference to Fig. 1-2. Consider the line labeled "solid-liquid," which represents for a typical pure metal the temperature at which melting occurs at various pressures. This line slopes upward away from the pressure axis. The typical metal contracts upon freezing. Hence, applying an increased pressure to the liquid can cause the metal to become solid, experiencing at the same time, an abrupt contraction in volume. Had the metal bismuth, which expands upon freezing, been selected as an example, the theorem of Le Châtelier would demand that the solid-liquid line be so drawn that the conversion of liquid to solid with pressure change would occur only with a reduction in pressure; that is, the line should slope upward *toward* the pressure axis.

A quantitative statement of the theorem of Le Châtelier is found in the Clausius-Clapeyron equation (see Appendix V). Referring again to Fig. 1-2, this equation leads to the further conclusion that each of the curves representing two-phase equilibrium must lie at such an angle that upon passing through the point of three-phase equilibrium, each would project into the region of the third phase. Thus, the solid-gas line must project into the liquid field, the liquid-gas curve into the solid field, and the solid-liquid curve into the gas field.

Other rules and aids of this type will be mentioned in subsequent chapters as opportunities for their application arise. Foremost among these is the second law of thermodynamics, which leads to rules governing the construction also of more complex phase diagrams such as are encountered in binary and ternary systems.

## Relationships between Alloy Constitution and Physical Properties

At the present stage of the development of the science of physical metallurgy no fundamental relationships have been established associating states of phase equilibrium with the physical and mechanical properties of alloys. Nevertheless, experience has shown that there are striking similarities in the structure and properties among alloy systems having similar phase diagrams.

One of the more successful correlations is that with the structural alterations that an alloy undergoes during temperature change, as in manufacturing operations. True equilibrium is, of course, rarely attained by metals and alloys in the course of ordinary manufacture and application. Rates of heating and cooling are usually too fast, times of heat treatment too short, and phase changes too sluggish for the ultimate equilibrium state to be reached. Any change that does occur must, however, constitute an *adjustment toward equilibrium*. Hence, the direction of change can be ascertained from the phase diagram, and a wealth of experience is available to indicate the probable degree of attainment of equilibrium under various circumstances.

## Advantages of Phase Diagrams

From what has been said in this introductory chapter it will be seen that the use of phase diagrams for recording phase changes in alloys offers three important advantages:

1. The conditions under which phase changes occur can be recorded simply and clearly for a large number of alloy compositions in relatively small space.

2. The existence of certain rules of construction greatly reduces the number of experimental observations necessary to determine the phase relationships that exist in a whole series of alloys.

3. The recognition of quasi relationships between the constitution of alloys and their structure and properties makes the phase diagram an invaluable guide in the control of metallurgical processes.

# CHAPTER 2

# UNARY SYSTEMS

Although little used as such in metallurgical work, the phase diagrams of one-component (unary[1]) systems are of importance because they provide the foundation upon which the diagrams of all multicomponent systems must be built. An understanding of the concepts of one-component diagrams is, therefore, prerequisite to a full appreciation of the meaning of the more complex phase diagrams. In Fig. 1-2 a typical diagram of this class has been presented.

Of the three externally controllable factors temperature, pressure, and composition, only the first two may be varied in a one-component system, because the composition must always be unity, i.e., the pure metal constitutes 100% of the material under consideration. The two remaining variables, temperature and pressure, may be represented on a two-dimensional diagram, where it is customary to plot the temperature (abscissa) along the horizontal axis and the pressure (ordinate) along the vertical axis (Fig. 2-1). This is called a $PT$, or *pressure-temperature*, *diagram*. The three states or phases —solid, liquid, and gas—are represented in the three correspondingly labeled areas. Equilibrium between any two phases occurs upon the mutual boundary of the areas concerned, and equilibrium among all three phases occurs at the intersection of the three boundaries.

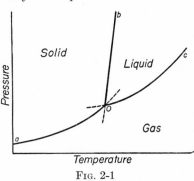

Fig. 2-1

---

[1] The use of Latin roots in the derivation of names for the systems of various numbers of components is preferred. This leads to the following terminology:

Unary = one component
Binary = two components
Ternary = three components
Quaternary = four components
Quinary = five components

Sexinary = six components
Septenary = seven components
Octanary = eight components
Nonary = nine components
Decinary = ten components

## Invariant Equilibrium

According to the phase rule three phases may coexist at equilibrium only at a single point on the diagram, point $O$ in Fig. 2-1. This is the *triple point*. It is also an *invariant*[1] *point* because there all the externally controllable factors are fixed, with definite values. If either the temperature or the pressure is caused to deviate from the stated values, one or two of the phases must disappear.

This may be seen more clearly in an example. Suppose that a sample of a pure metal, such as zinc, is confined in a chamber made of a substance toward which the metal is inert. The chamber is sealed with a tightly fitting piston by means of which the enclosed volume may be controlled. Now, if the piston is lifted to provide some vacant space around the metal sample and the whole apparatus is heated until a portion of the metal has melted but is then held at constant temperature without further introduction of heat, the pure metal will come to equilibrium with solid, liquid, and vapor all present together. A slight increase in the temperature would cause the inward flow of heat until the solid portion disappeared by melting; a decrease would cause the liquid to disappear by freezing. If, in another experiment, an attempt were made to raise the pressure by lowering the piston, it would be found that no rise in pressure could be observed until the vapor was entirely condensed, after which further lowering of the piston would result in a pressure increase. Thus, any permanent change in either the temperature or the pressure reduces the number of phases present.

In the above example, particular note should be made of the observation that no increase in the pressure occurred until the gas phase had disappeared; a partial lowering of the piston caused some of the vapor to condense, but there was no change in the number of phases present or in the pressure (or temperature) of the system. Similarly, if at constant volume enough heat had been introduced to melt a part, but not all, of

---

[1] Terms used to designate the number of degrees of freedom are preferably derived from Latin roots, as follows:

$$
\begin{array}{rl}
\text{Invariant} & = \text{no degrees of freedom} \\
\text{Univariant} & = \text{one degree of freedom} \\
\text{Bivariant} & = \text{two degrees of freedom} \\
\text{Tervariant} & = \text{three degrees of freedom} \\
\text{Quadrivariant} & = \text{four degrees of freedom} \\
\text{Quinquevariant} & = \text{five degrees of freedom} \\
\text{Sexivariant} & = \text{six degrees of freedom} \\
\text{Septevariant} & = \text{seven degrees of freedom} \\
\text{Octavariant} & = \text{eight degrees of freedom} \\
\text{Nonavariant} & = \text{nine degrees of freedom} \\
\text{Decivariant} & = \text{ten degrees of freedom}
\end{array}
$$

the solid present, the temperature, pressure, and phase count would have remained constant. This serves to illustrate the important fact that the *phase rule is not concerned with the total or relative quantities of the phases present or with their related volumes and heat contents;* it is concerned only with the number of phases present and with their temperature, pressure, and composition.

## Univariant Equilibrium

Equilibrium between two phases in a one-component system occurs with one degree of freedom and is called *univariant equilibrium.*

$$P + F = C + 2$$
$$2 + 1 = 1 + 2$$

This means that either the pressure or the temperature may be freely selected, but not both. Once a pressure is chosen at which two phases are to be held in equilibrium, there will be only one temperature that will satisfy the requirements, or conversely. Since it is permissible to select the pressure (or temperature) willfully, whereupon it is found that the other variable, the temperature (or pressure), has a corresponding dependent value, it is apparent that equilibria between any two phases must be represented on the one-component diagram by a line, or curve.

Three curves of univariant equilibrium issue from the triple point in Fig. 2-1 and are called *triple curves.* They represent the equilibria between the solid phase and gas (line $aO$), between liquid and gas (line $Oc$), and between the solid and liquid (line $Ob$). There are always three, and only three, triple curves issuing from each triple point, because the three phases represented at the triple point can be paired in only three ways.

The *vaporization, or liquid-gas, curve* extends from the triple point to a *critical point* at $c$. There the curve ends; at higher temperature the liquid and gas phases are indistinguishable. If a liquid partially filling a closed container is heated, the pressure will rise in accordance with the demands of line $Oc$ and the vapor over the liquid will gradually become denser as the pressure increases. When the temperature and pressure of the critical point are reached, the densities of liquid and gas will be identical, the meniscus (interface between liquid and gas) will have vanished, and the gas and liquid phases will be one. The fact that this two-phase equilibrium can exist over a limited temperature and pressure range only is in no way contradictory to the exercise of a degree of freedom. The "free" choice of pressure or temperature must obviously remain within the ranges where both phases concerned are themselves stable and distinguishable.

Along the triple curve $aO$ the solid and gas phases are coexistent. This is the *sublimation curve*. It proceeds from the triple point always to lower temperature and lower pressure and may be presumed to end at the absolute zero of temperature at zero pressure. The *melting curve Ob*, along which solid and liquid are in equilibrium, has no known terminus (except in those cases where a new type of crystal appears at high pressure, when the curve ends upon another triple point representing the co-existence of liquid and two kinds of solid, Fig. 2-2). If the metal expands upon melting, this curve inclines upward to the right, in accordance with

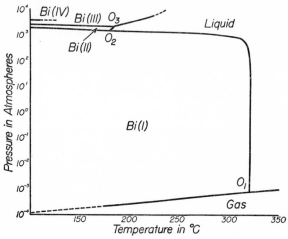

Fig. 2-2. Pressure-temperature diagram of bismuth. The abruptness with which the melting curve bends toward lower pressure at higher pressure is the result of having plotted the pressure upon a logarithmic scale. Three triple points are shown: Bi$_I$-*L-G*, Bi$_I$-Bi$_{II}$-*L*, and Bi$_{III}$-Bi$_{III}$-*L*. The full range of existence of the fourth solid phase Bi$_{IV}$ has not been determined. (*After P. W. Bridgman.*)

the requirements of the theorem of Le Châtelier, or if the metal contracts during melting (bismuth and gallium, for example), the melting curve inclines upward to the left.

## Metastable Univariant Equilibria

Each of the triple curves in Fig. 2-1 has been drawn with a dotted portion extending beyond the triple point. This construction, which is normally omitted, serves to demonstrate that the diagram has been drawn in accordance with the demands of the Clausius-Clapeyron equation. As a practical guide in drawing PT (pressure-temperature) diagrams of this kind, it is useful to remember that either the extension of each triple curve should lie within the field of the phase not represented on that triple curve or, equivalently, the angle between adjacent triple curves at their intersection with the triple point should never exceed 180° of arc.

The extensions of the triple curves do, in fact, have physical significance in certain cases. Suppose, for example, that a body of liquid, in equilibrium with its vapor, is cooled through the triple point and undercooling occurs, i.e., the solid phase fails to appear. When this happens, the vapor pressure, in equilibrium with the undercooled liquid, is that shown by the dotted extension of the liquid-gas curve $Oc$. The system is said to be in a state of *metastable equilibrium*. A similar metastability may exist in the event that the solid fails to melt when heated through the triple point. Here the equilibrium is represented upon the extension of the solid-gas curve. In each case the metastable portion of the curve lies at higher pressure than does the curve of stable equilibrium at the same temperature. This is a general rule, namely: *the vapor pressure of a metastable phase is, at a given temperature, always greater than that of the stable phase.*

## Bivariant Equilibrium

If the temperature and pressure are each fixed arbitrarily and independently, this condition corresponds to the exercise of two degrees of freedom, *bivariant equilibrium*, and only one phase can exist in a stable state.

$$P + F = C + 2$$
$$1 + 2 = 1 + 2$$

Points on the phase diagrams designated by an independent selection of pressure and temperature values will occupy an area whereby it is seen that the areas between the triple curves in Fig. 2-1 must represent the existence of single phases.

## Interpretation of the One-component Phase Diagram

A list of estimated triple points of some of the more common metals is given in Table 1. For the majority of the metals the disposition of the triple curves is such that the triple point lies far below atmospheric pressure and the critical point well above atmospheric pressure, as in Fig. 2-3, where the pressure at $P_2$ is considered to be 1 atmosphere.

The behavior of such a metal during heating under a constant pressure of 1 atmosphere (isobaric heating) may be predicted by following the line $P_2$ from left to right. At the minimum temperature (left end, point 1), the solid phase is surrounded by a gas phase exhibiting a partial pressure $a$ of metal vapor plus a much larger partial pressure $1a$ of air, to make a total pressure of 1 atmosphere. If the system is enclosed, this equilibrium will be maintained; if open to the free passage of air, the metal vapor will drift away and the partial pressure of the metal will be maintained only to the extent that sublimation from the metal surface can keep pace with the loss of metal vapor.[1] As the temperature is increased, the solid phase

TABLE 1. TRIPLE POINTS OF COMMON METALS (COMPUTED)

| Metal | Temperature, °C | Pressure, atmospheres |
|---|---|---|
| Arsenic.............. | 814 | 36 |
| Barium.............. | 704 | 0.001 |
| Calcium............. | 850 | 0.0001 |
| Copper............. | 1083 | 0.00000078 |
| Iron ($\delta$)............. | 1535 | 0.00005 |
| Lead............... | 327 | 0.0000001 |
| Manganese.......... | 1240 | 0.001 |
| Mercury............ | −38.87 | 0.0000000013 |
| Nickel............. | 1455 | 0.0001 |
| Platinum........... | 1773 | 0.000001 |
| Silver.............. | 960 | 0.0001 |
| Strontium........... | 770 | 0.0001 |
| Zinc............... | 419 | 0.05 |

remains unchanged, while the partial pressure of the metal vapor increases along the curve $aO$ to that of the triple point. At nearly the same temperature the metal melts isothermally (point 2). The physical properties,

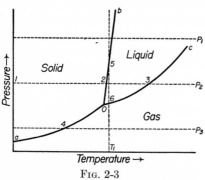

Temperature →

FIG. 2-3

such as density and electrical conductivity, as well as the mechanical properties suffer discontinuous change. Above the melting temperature the partial pressure of the metal vapor continues to rise, but more slowly at first, until the vaporization curve $Oc$ intersects atmospheric pressure (point 3). This is the boiling point; at higher temperatures only metal vapor can exist. During cooling, the same changes occur in reverse order: condensation of gas to liquid (point 3), freezing of the liquid (point 2), and a reduction of the partial pressure of the gas phase along the corresponding triple curves.

The metal arsenic has its triple point at a pressure of 36 atmospheres.

---

[1] With most metals the vapor pressure is so low at room temperature as to be entirely negligible; indeed, it has been calculated that the vapor pressure of tungsten would be equivalent to one atom in the entire solar system. In a few cases, however, the pressure is significant.

In this case, 1 atmosphere of pressure lies far below the triple point, as at $P_3$ in Fig. 2-3. Upon heating, the partial pressure of arsenic vapor reaches 1 atmosphere at 350°C (point 4). Here, given time, the solid phase will be converted directly to gas; i.e., it will sublime. The rate of sublimation is low, however, and it is possible to heat arsenic to and above the temperature of the triple point before it has evaporated. When this is done, melting occurs near the temperature of the triple point.

Another conceivable case is that in which 1 atmosphere of pressure lies above the critical point $c$, as at $P_1$. Here, conditions are similar to those described for the case where 1 atmosphere lay at $P_2$, except that no boiling point will be observed. No examples of this kind have as yet been identified among the metals.

Changes accompanying *isothermal* (constant temperature) pressure change follow a similar pattern. Consider the case represented in Fig. 2-3, when the temperature is maintained at $T_1$ and the pressure is gradually reduced from its maximum value. At point 5 melting takes place. (Actually this is a very improbable event, because the slope of the melting curve is usually so slight that the liquid-solid change cannot be observed within the ordinarily attainable range of pressure.) Further decrease in pressure through point 6 would result in the ultimate vaporization of the metal. The latter condition is sometimes observed in vacuum systems.

## Allotropy

Some metals are capable of existing in more than one crystalline form. The metal is then said to be *allotropic*, and the individual crystal varieties are known as *allotropes* or *allotropic modifications*. The transformation from one allotrope to another is *allotropic transformation*. This may occur either with pressure change or with temperature change. Bismuth, which exists in only one solid form at normal pressures, has been found to be capable of stable existence in at least two other crystalline forms at very high pressures (Fig. 2-2). Iron, on the other hand, passes through two allotropic transformations when it is heated from room temperature to its melting point (Fig. 2-4).

The PT diagram of an allotropic metal always has more than one triple point, because no more than three phases may be brought into simultaneous equilibrium in a one-component system. Only one of the triple points can involve both liquid and gas; all others represent equilibrium between two crystalline phases and gas, two crystalline phases and liquid, or three crystalline phases. It is noteworthy that successive triple points involving the gas phase (Fig. 2-4) must occur *at ascending temperature and pressure*. No equivalent generalization is possible in the case of triple points involving the liquid phase.

### Concerning the Gas Phase in Higher-order Systems

In the treatment of binary, ternary, and higher-order systems immediately following, discussion of equilibria involving the gas phase will generally be omitted. This is done in order to avoid complications which might obscure the principal subject matter. It should be understood, however, that there is always a gas phase present, except when a liquid or solid material is enclosed within and completely fills a sealed container.

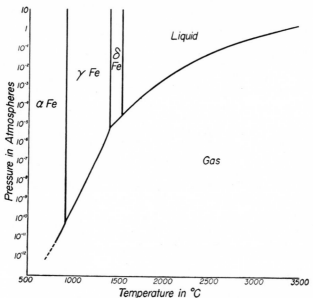

Fig. 2-4. Estimated pressure-temperature diagram of iron.

Since it is customary to handle metals in the open, where the gas phase can exist, the representation of the corresponding phase equilibria without reference to the gas phase constitutes an incomplete expression of the state of equilibrium. For most purposes this omission is inconsequential. Where it is significant, the use of diagrams involving the pressure variable is required (see Chap. 20).

### PRACTICE PROBLEMS

**1.** Draw upon the same coordinate scale estimated PT diagrams for: arsenic, zinc, manganese, silver, lead, copper, and mercury.

**2.** The metal cobalt freezes at 1478°C to the face-centered cubic crystal form, which in turn transforms at 420°C to the close-packed hexagonal crystal form; its vapor pressure approximates that of its sister metals nickel and iron. Draw an estimated PT diagram for cobalt.

**3.** A metal is sealed in a completely inert and pressure-tight container. Initially the metal half fills the container, the balance of the space being filled with an inert gas at a pressure of 1 atmosphere. The phase diagram is similar to that given in Fig. 2-3. What changes will occur within the container as this system is heated?

**4.** A piece of pure iron is held in a continuously pumped vacuum of 0.000001 atmosphere and is heated, during a period of 10 min, to a temperature of 1550°C. What changes will the iron undergo? What difference in behavior should be noticed if the heating rate were reduced to one-thousandth that of the first experiment?

# CHAPTER 3

# BINARY ISOMORPHOUS SYSTEMS

A *binary isomorphous* system is one in which two metals are mutually soluble in all proportions and in all states, the solid, liquid, and gas. The concept of liquid solution is familiar, and it is easy also to conceive of gaseous solution, because gases mix in all proportions. In solid solutions[1] the solute metal simply enters and becomes a part of the crystalline solvent, without altering its basic structure. There are two kinds of solid solutions: (1) *substitutional solid solutions*, in which the solute atom occupies, in the solvent crystal, a position belonging to one of the atoms of the solvent metal, and (2) *interstitial solid solution*, in which the solute atom enters one of the vacant spaces between atoms in the lattice of the solvent crystal without displacing a solvent atom. Isomorphous solid solutions are always of the substitutional kind. Both of the metals involved must have the same type of crystal structure, because it must be possible to replace, progressively, all the atoms of the initial solvent with atoms of the original solute without causing a change in crystal structure (i.e., a phase change).

In order to produce a complete phase diagram of a binary alloy system in which the three externally variable factors, pressure, temperature, and composition, are all represented, it is necessary to resort to the use of a three-dimensional space model as in Fig. 3-1. This is a *PTX diagram* (pressure-temperature-composition). Pressure is measured vertically, temperature horizontally to the rear, and composition horizontally across the front of the model. The composition is usually expressed in weight percentage as described in Chap. 1. The right side of the figure represents pure metal $B$ and is the PT diagram of that metal; the left side, in like manner, is the PT diagram of pure metal $A$ (compare with Fig. 2-1).

Within the space model there are three pairs of curved surfaces which serve to join the corresponding triple curves of the two pure metals. One of these pairs of surfaces, which joins the sublimation curves of $A$ and $B$, intersects the front surface of the model in two curved lines $a_A a_B$, showing that the two surfaces enclose a space of lenticular cross section in which

---

[1] In some of the older literature the term "mixed crystals," a direct translation of the German *Mischkristalle*, is used in place of the term "solid solution."

the solid (solution) and gas (solution) phases coexist. Another such pair $b_A b_B$ joining the melting curves encloses a similarly shaped space in which the liquid (solution) and solid (solution) phases coexist. The third pair joins the vaporization curves and encloses a space where liquid and gas coexist. Between the two-phase regions, three large spaces represent the conditions of individual stability of the three phases of the system. These

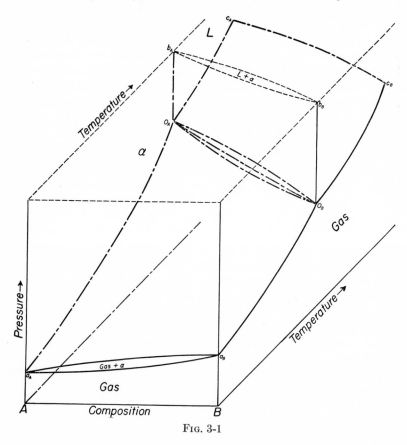

FIG. 3-1

are the gas-solution region in the space along the bottom of the diagram, the solid-solution region in the upper front space, and the liquid-solution region in the upper rear portion of the model. Attention is directed to the condition that the one-phase regions are separated from each other by two-phase regions everywhere except at the sides of the diagram, where the one-phase regions contact one another along the triple curves of the pure metals.

Although the space diagram must extend upward indefinitely, the diagram in Fig. 3-1 has been arbitrarily cut off by a horizontal (isobaric) plane. If the resulting cross section is considered to lie at 1 atmosphere of pressure, it will represent the phase relationships in all alloys composed of the metals $A$ and $B$ at normal pressure. This, then, is the usual *TX diagram* (temperature-composition diagram) of the $AB$ alloy system (see Fig.

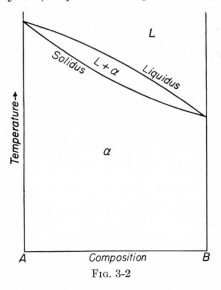

3-2 and compare with Fig. 1-1). Further discussion of the space diagram will be deferred to a later chapter. For the present it will be sufficient to appreciate that the TX diagrams, that are about to be discussed, are all (horizontal) sections taken arbitrarily through the PTX space diagram at a constant pressure of 1 atmosphere.

Fig. 3-2

### Applications of the Phase Rule

In applying the phase rule to the binary TX diagram it is important to remember that this is an isobaric section in which the pressure has been selected by free and arbitrary choice and that the act of selecting a pressure constitutes the exercise of one degree of freedom. A one-phase region, such as the $\alpha$ or the $L$ region, is tervariant; i.e., there are three degrees of freedom.

$$P + F = C + 2$$
$$1 + 3 = 2 + 2$$

Any point in a one-phase region is fixed by selecting first, a pressure of 1 atmosphere, which places the point somewhere upon the isobaric section under consideration; second, a temperature, which locates the vertical position of the point; and finally, a composition, which locates its lateral position. Three degrees of freedom have been utilized by the independent choice of pressure, temperature, and composition.

When two phases are coexistent, as in the $L + \alpha$ region of the diagram, it is evident that they must be at the same pressure and temperature, for physically they must lie side by side in the same environment, but it is not necessary that they have the same composition. The phase rule requires that there be two degrees of freedom in a two-phase two-component mixture.

$$P + F = C + 2$$
$$2 + 2 = 2 + 2$$

The pressure has been selected in establishing the isobaric section, and only one degree of freedom remains. If this degree of freedom is employed in establishing the temperature, it follows that the compositions of the two phases must be fixed thereby.

This is illustrated in Fig. 3-3, where a horizontal line (dashed) has been drawn across the $L + \alpha$ region of the phase diagram at the arbitrarily selected temperature $T_1$. Every point on the horizontal line except the ends lies within the $L + \alpha$ field and must represent the coexistence of liquid and solid solution. The left end, at $\alpha_1$, lies upon the solidus, where the alloy must be completely solid, and the right end lies upon the liquidus, at $L_1$, where the alloy must be entirely molten. If, as was said above, the composition of the solid phase must be the same in all mixtures with liquid at this temperature, it is apparent that the composition so

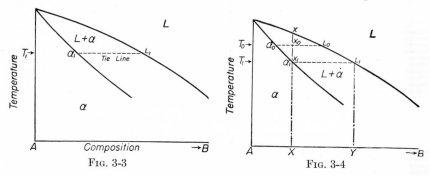

Fig. 3-3          Fig. 3-4

designated must be that at $\alpha_1$, because this is the only composition that the solid could assume as the quantity of liquid approaches zero. Likewise, the composition of the liquid phase must be that at $L_1$, where the quantity of solid reaches zero. Every mixture of $L$ and $\alpha$ at this temperature $T_1$ must, then, be composed of various quantities of $\alpha_1$ and $L_1$.

The horizontal line connecting $\alpha_1$ and $L_1$ is known as a *tie-line*[1] because it joins the composition points of the *conjugate phases* that coexist at a designated temperature and pressure.

Exactly the same conclusions would have been reached if, instead of using the one remaining degree of freedom to designate a temperature, the choice of a composition had been elected. Suppose that it had been required that the $\alpha$ phase contain $X\%$ $B$. This should result in the simultaneous establishment of fixed values of the temperature and of the composition of the liquid phase. According to the stipulation that the solid phase must have the composition $X$, the point representing the solid must lie somewhere on the line $Xx$ in Fig. 3-4. Let the conditions be tested at some randomly selected temperature $T_0$. It will then appear, by

[1] In some of the literature the German word *konode* is used in place of the English equivalent *tie-line*.

repeating the argument cited in the previous example, that the only possible solid composition at this temperature is $\alpha_0$ and not $x_0$. Similar results will be obtained at any other temperature except $T_1$; here $\alpha_1$ and $x_1$ are identical points, and all requirements are satisfied. By establishing the composition of the solid phase at $X$, the temperature of the equilibrium with liquid has been fixed at $T_1$ and the composition of the liquid phase has been fixed at $Y$. Two degrees of freedom, namely, pressure and composition, have been exercised, as predicted by the phase rule.

A further significance is now seen to reside in the liquidus and solidus curves. The solidus designates for each temperature the composition of the solid phase that may exist in equilibrium with liquid, and the liquidus designates for the same temperature the composition of the liquid phase

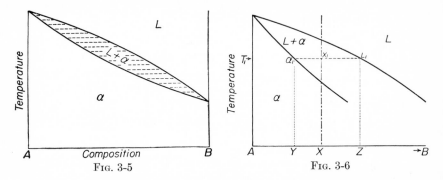

FIG. 3-5          FIG. 3-6

in equilibrium with solid. The $L + \alpha$ region may be thought of as being composed of an infinite number of tie-lines connecting the conjugate solid and liquid compositions (Fig. 3-5) at all temperatures within the limits of the field. It is customary to omit the tie-lines from binary diagrams, because their location is self-evident.

## The Lever Principle

Further consideration of the nature of the tie-line reveals that it may be used to compute the relative quantities of two phases present at equilibrium in an alloy of stated gross composition at a given temperature. It has been pointed out that at temperature $T_1$ in Fig. 3-6, the alloy of composition $Y$ is composed wholly of solid ($\alpha_1$) while the alloy of composition $Z$ is composed wholly of liquid ($L_1$); all intermediate compositions at this temperature must be composed of mixtures of various proportions of $\alpha_1$ and $L_1$. Since the compositions of the two phases are the same in all mixtures, there must be a linear relationship obtaining between the alloy composition and the proportions of the phases present at fixed temperature; i.e., the percentage of liquid must increase linearly from 0 at composition $Y$ to 100 at composition $Z$. The tie-line can, therefore, be treated as a lever with its fulcrum at the gross composition of the alloy under consideration and the total weight of each of the two phases impinging upon the lever at the respective ends, where their compositions are represented.

In Fig. 3-6, the lever is the tie-line $\alpha_1 L_1$, balanced upon its fulcrum at $x_1$, the gross composition of the alloy under consideration. The length $x_1 L_1$ will then be proportional to the total, or relative, weight of the solid phase of composition $\alpha_1$, while the length $\alpha_1 x_1$ will be proportional to the total, or relative, weight of the liquid phase $L_1$. If the whole length of the tie-line $\alpha_1 L_1$ is taken to represent 100% (or the total of the two phases present in alloy $X$ at temperature $T_1$), then the length $x_1 L_1$ will be the percentage of solid and $\alpha_1 x_1$ the percentage of liquid. This is the *lever principle*, which may be conveniently expressed in mathematical form as follows:

$$\%\alpha_1 = \frac{x_1 L_1}{\alpha_1 L_1} \times 100$$

$$\%L_1 = \frac{\alpha_1 x_1}{\alpha_1 L_1} \times 100$$

### Equilibrium Freezing of a Solid-solution Alloy

The phase changes that accompany freezing under equilibrium conditions in a given alloy can be read from the phase diagram by observing the phases that are present and their compositions at each successively lower temperature. Let $X$, Fig. 3-7, represent the gross composition of the alloy in which freezing behavior is to be considered. Observations will begin at temperature $T_0$, where the alloy is entirely molten, point $x_0$. No phase change occurs upon cooling until the temperature $T_1$ is reached, whereupon freezing begins with the initial deposition of crystals of com-

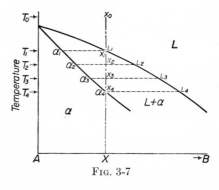

Fig. 3-7

position $\alpha_1$, which is the only composition of solid that can be in equilibrium with liquid having the gross composition $X$, point $L_1$. The separation of crystals of $\alpha_1$ takes from the liquid a disproportionately large amount of the $A$ component, causing the liquid to become richer in the $B$ component. Thus, when a lower temperature $T_2$ is reached, the composition of the liquid has shifted to $L_2$. The only solid that can remain in equilibrium with $L_2$ is $\alpha_2$. Hence, the composition of the solid must also have moved toward higher $B$ content. This can happen only by the preferential absorption of $B$ from the liquid and the diffusion of the $B$ metal into the solid formed previously at higher temperature. When equilibrium is established at temperature $T_2$, the relative quantities of the solid and liquid phases may be computed by applying the lever principle:

$$\%\alpha_2 = \frac{x_2 L_2}{\alpha_2 L_2} \times 100 \approx 50\%^*$$

$$\%L_2 = \frac{\alpha_2 x_2}{\alpha_2 L_2} \times 100 \approx 50\%^*$$

Growth of the solid at the expense of the liquid continues, with the compositions of both phases shifting toward greater $B$ content as the temperature falls. At $T_3$ the solid phase will constitute about three-quarters of the material present and its composition will lie at $\alpha_3$. Finally, at $T_4$, the liquid will disappear:

$$\%\alpha_4 = \frac{x_4 L_4}{\alpha_4 L_4} \times 100 = 100\%$$

The alloy is now completely frozen, and the composition of the solid phase has arrived at the gross composition of the alloy $X$.

### Origin of Coring

The conditions of "equilibrium" freezing are not realizable in practice because of the very long time required at each decrement of temperature to readjust the compositions of the two phases. This adjustment can be accomplished only by the diffusion of $A$ and $B$ atoms in both phases, together with the exchange of atoms across the liquid-solid interface. Since diffusion proceeds slowly, especially in the solid state, true equilibrium cannot be maintained during cooling except at an infinitely slow rate.

Certain departures from equilibrium freezing are, therefore, to be expected when ordinary cooling rates are used. In the example about to be described it will be assumed that homogeneous equilibrium is maintained in the liquid phase,[1] but not in the solid phase, where the rate of diffusion is much lower. Referring again to the alloy of composition $X$, Fig. 3-8, freezing begins with the deposition of crystals of $\alpha_1$ when the temperature $T_1$ is reached. At $T_2$ the liquid composition has shifted to $L_2$, and the

---

\* The inclusion of an approximate percentage here and elsewhere is intended to assist the reader in comparing the relative quantities of phases represented on the tie-lines.

[1] It is scarcely likely that equilibrium could be maintained in the liquid phase during freezing, but this assumption simplifies the argument without vitiating its conclusions.

solid forming at that temperature will have the composition $\alpha_2$. But the solid which had formed at temperatures above $T_2$ has not changed altogether from its initial compositions, and the average composition of all the solid formed up to this point will be represented at $\alpha_2'$. As the temperature drops to $T_3$, the average solid composition $\alpha_3'$ is seen to depart more and more from the equilibrium composition $\alpha_3$. Under ideal conditions freezing should be completed at $T_4$. Here, however, the average solid composition has not reached the gross composition

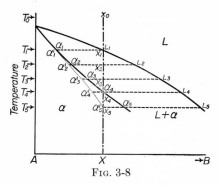

Fig. 3-8

of the alloy $X$, and some liquid must remain. The relative quantities may be read from the tie line at $T_4$:

$$\% \text{ solid} = \frac{x_4 L_4}{\alpha_4' L_4} \times 100 \approx 85\%$$

$$\% \text{ liquid} = \frac{\alpha_4' x_4}{\alpha_4' L_4} \times 100 \approx 15\%$$

Freezing continues, therefore, to lower temperature, and the solid now being deposited is richer in the $B$ component than the original composition of the alloy $X$. Finally, at $T_5$ the average solid composition $\alpha_5'$ coincides with the gross composition at $x_5$ and the alloy is completely solidified.[1]

Nonequilibrium freezing is characterized by an increased temperature range over which liquid and solid are present and by a composition range remaining in the solidified alloy. If the rate of cooling is decreased, the displacement of the average solid compositions from the solidus will be less; likewise, if the diffusion rate in the solid state had been greater, the departures from equilibrium would have been less.

Undercooling normally precedes the initiation of freezing in alloys; i.e., the liquid cools below $L_1$ before freezing starts. When this happens, the first solid may have a composition nearer to the gross composition of the alloy than is $\alpha_1$. However, the release of latent heat of crystallization usually elevates the temperature again (recalescence) so that a modest

[1] For more detailed discussions of the theory of natural freezing, see G. H. Gulliver, The Quantitative Effect of Rapid Cooling upon the Constitution of Binary Alloys, *J. Inst. Metals*, **9**:120–153 (1913), **11**:252–272 (1914), **13**:263–291 (1915); J. F. Russell, The Interpretation of Thermal Curves and Some Applications to Ferrous Alloys, *J. Iron Steel Inst.*, **139**:147–176 (1939); E. Scheuer, Zum Kornseigerungsproblem, *Z. Metallkunde*, **23**:237–241 (1931).

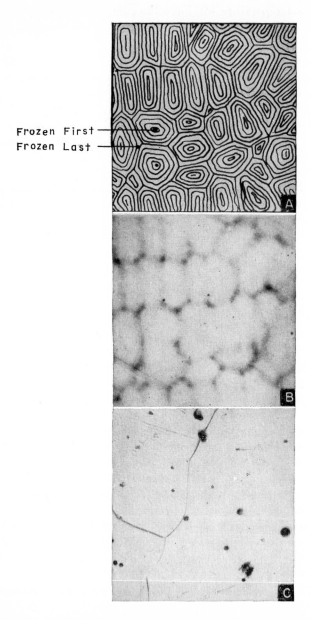

Frozen First

Frozen Last

FIG. 3-9. Illustrates coring in a cast 70% Ni + 30% Cu alloy. Nickel-rich areas are light gray, copper-rich areas are dark gray in *B*, magnification 100. The order of freezing of the cored structure is indicated upon the tracing *A*. Long heat treatment at a high temperature produces a uniform distribution of the copper and nickel as shown in photomicrograph *C*, magnification 100.

degree of initial undercooling may be expected to have but a minor influence upon the ultimate path of freezing.

The solidification of an alloy begins with the appearance of small solid particles (nuclei) which, barring undercooling, contain more of the higher melting component metal than any subsequently formed solid. Upon the nuclear crystallites successive layers of the solid phase are deposited, each layer being a little richer than its predecessor in the low-melting component. The final solid is composed, then, of a "cored" structure, in which each unit has a high-melting central portion surrounded by lower melting material (see Fig. 3-9$A$). The majority of metals freeze by the formation of treelike growths, called dendrites, whose branches become thicker as freezing proceeds until all the space is filled with solid and a continuous crystal has been developed. Hence, the cored structure in most metals has the geometry of the dendrite. This condition is sometimes referred to as *dendritic segregation*.

Seen in cross section under the microscope, the sectioned branches of the dendrite appear in the more or less regular pattern of the rounded figures depicted in Fig. 3-9$A$. A photomicrograph of a cored copper-nickel (30-70) alloy (see Fig. 1-1) is presented in Fig. 3-9$B$. This picture was obtained by oxidizing the polished surface of the sample in the air; the more easily oxidized, lower-melting, copper-rich regions of the alloy were darkened by this procedure, while the higher-melting, tarnish-resistant, nickel-rich cores were left bright.

## Homogenization

For some uses the cored cast structure is objectionable and must be eliminated by a suitable heat treatment. The equalization, or *homogenization*, of the composition of the cast alloy, by eliminating coring, occurs spontaneously through the action of diffusion under favorable conditions. Diffusion in the solid state is most rapid at a temperature just below that of melting, and in most alloys its rate is approximately halved with each 50°C lowering of the temperature. Except with metals of very low melting point, diffusion is scarcely detectable at room temperature.

The choice of a suitable heat-treating temperature for homogenization is limited on the high side by the onset of melting if the solidus curve is crossed and on the low side by the marked deceleration of the process as the temperature is decreased. Allowance must be made also for the fact that some of the solid, which is richer in the low-melting component than the average composition of the alloy ($\alpha_5$ in Fig. 3-8), may begin to melt below the equilibrium solidus temperature of the alloy. The equilibrium solidus may be approached more closely without inducing melting if the cast alloy is heated slowly than if it is quickly heated to a high temperature, because time is thereby furnished for the lowest melting solid to be eliminated.

After homogenization, the dendritic cored structure is no longer visible in the microstructure (Fig. 3-9C). It should be noted that homogenization is an adjustment toward a state of *homogeneous equilibrium* not involving a *heterogeneous phase change*. Therefore, the phase diagram does not predict the change and is useful, in this case, only in the selection of a suitable heat-treating temperature.

### Liquation

Where an alloy has been heated only slightly above the solidus temperature, so that melting has just begun, it is said to have been *liquated*, or *burned*. Melting usually begins at the grain boundaries and interdendritic regions of the alloy, forming a thin film of liquid that separates each crystal of the solid from its neighbor (Fig. 3-11A). The strength of the alloy is destroyed almost immediately with the onset of melting. Large or complex castings will sag out of shape, and if disturbed while liquated, the material may crack and fall to pieces. Alloys with a long temperature interval between liquidus and solidus are often characterized as being "hot short" because their strength and ductility may be destroyed by liquation at a temperature far below that at which the onset of melting is otherwise evident. As has been pointed out above, melting (liquation) may be encountered below the solidus temperature if the alloy is cored.

Liquation may also have an effect upon the physical properties and microstructure of the alloy after it has been returned to room temperature, because a special type of coring is developed at the grain boundaries

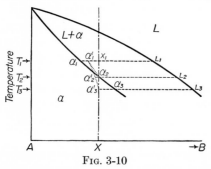

FIG. 3-10

and at other localities where liquid has appeared. The reason for this can be seen by reference to Fig. 3-10. Assume that the alloy $X$ has been homogenized, melting will begin when the temperature $T_2$ is reached, the liquid having the composition $L_2$. When temperature $T_1$ is reached, the liquid composition will be $L_1$, but the composition of the solid phase will not have had time to reach its equilibrium value at $\alpha_1$; instead the average solid composition will lie at some intermediate point $\alpha_1'$. For this reason the quantity of liquid formed will be less than that to be expected at equilibrium:

$$\%L_1 = \frac{\alpha_1' x_1}{\alpha_1 L_1} \times 100 \approx 5\%$$

If the material is now rapidly cooled, freezing will occur in the normal manner, with a layer of $\alpha_1$ being deposited first on existing solid grains, followed by layers of increasing $B$ content up to $\alpha_3$, at $T_3$, where the supply of liquid is exhausted. Thus, coring occurs in the melted zone, coincident with the grain boundaries (Fig. 3-11). A photomicrograph of this effect in the copper-nickel alloy used for previous demonstrations is presented in Fig. 3-11$B$; compare the breadth of the grain boundary zone in the liquated sample with the narrow grain boundary in the homogenized

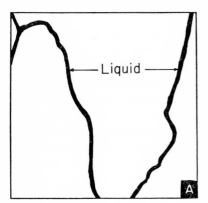 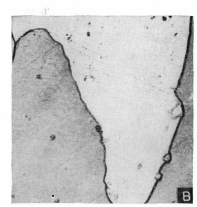

Fig. 3-11. Sketch in $A$ indicates that portion of the metal which was melted. Upon subsequent freezing, the center of the melted zone became copper-rich, and this has caused the grain boundaries in $B$ to appear as broad dark lines, magnification 100. Compare with the grain boundaries in Fig 3-9$C$.

sample (Fig. 3-9$C$). The effects of liquation can usually be eliminated by the same kind of heat treatment as that employed for the homogenization of cast alloys.

## Annealing

Solid-solution alloys are commonly subjected to annealing heat treatments to eliminate or to modify the effects of plastic forming at room temperature and to control grain size. The alterations in the metal that are affected thereby involve no phase changes, so that the phase diagram cannot, in general, be used to predict what will occur. It can sometimes be of assistance, nevertheless, in selecting suitable annealing temperatures. As a very rough approximation it can be said that a temperature (centigrade or Fahrenheit) of approximately one-half of the melting point of the alloy will usually produce a full anneal and a temperature of about one-quarter of the melting point will usually serve to relieve internal stresses. There are important exceptions to this statement, particularly among the metals of low melting point.

## Isothermal Diffusion

If pieces of the two component metals of an isomorphous system are brought into intimate contact and are maintained for some time at a temperature where diffusion is appreciable, alloying will take place in the solid state. During the early stages the composition will vary from pure metal $A$ at one side of the composite sample to pure metal $B$ at the other side. Much later, the composition will become uniform throughout and will be the same as though the two metals were alloyed by melting them together.

No phase change is involved in the above examples. If, however, the temperature of diffusion lies above the melting point of one of the metals, a phase change does occur. Suppose that temperature $T_1$ (Fig. 3-6) is chosen for the experiment and the pieces of metals $A$ and $B$ are of such relative size that together they would have the composition $X$. Metal $B$ is melted, and the piece of metal $A$ is immersed in the melt. Soon the composition of the solid will vary from pure $A$ at the center to $Y$ at the surface, while the liquid composition will vary from $Z$ at the surface of contact with the solid to pure $B$ at remote locations. After a time the solid will all reach the composition $Y$ and the liquid will reach the composition $Z$. The two phases will then be in stable equilibrium. If, on the other hand, the quantity of $A$ had been relatively larger, so that the gross composition had lain to the left of $Y$, the liquid phase would have disappeared entirely as equilibrium was approached, all of the $B$ component being used to form the solid solution. Or if the gross composition had lain to the right of $Z$, the solid phase would have dissolved completely in the melt.

## Physical Properties of Isomorphous Alloys

The physical properties of alloys can be divided into two groups, namely, those which are structure sensitive and those which are structure insensitive. *Structure-insensitive properties*, such as density, change only with the externally controllable variables, temperature, pressure, and composition. In the ideal case the change is linear (see, for example, the density vs. composition curve in Fig. 3-12), but both positive and negative deviations are encountered in nonideal isomorphous alloys. The *structure-sensitive properties*, such as hardness, tensile strength, and electrical conductivity, vary with the grain size, degree of coring, and other structural conditions. If these conditions are held constant and the composition is varied (Fig. 3-12), it is found that the change in properties is not linear but passes through a maximum or minimum. The change in properties is most rapid with the first addition of alloying element to either of the component metals and continues at a decelerating pace until reversal occurs at the maximum or minimum. Although the absolute values of the several properties differ widely from system to system, the forms of the curves are generally the same in all binary isomorphous alloy systems.

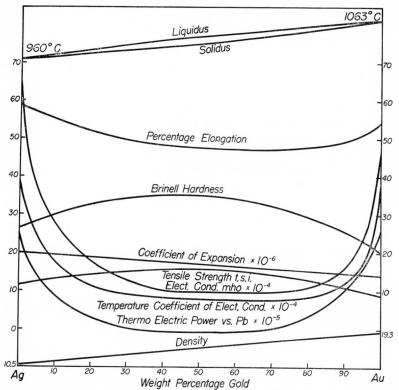

Fig. 3-12. Properties of annealed silver-gold alloys at room temperature. (*From Broniewski and Wesolowski.*)

## Minima and Maxima

There are some binary isomorphous systems in which the liquidus and solidus do not descend continuously from the melting temperature of one metal to that of the other but, instead, pass through a minimum temperature, which lies below the melting points of both components (Fig. 3-13). When this occurs, the *liquidus and solidus meet tangentially** at the minimum point.* That these lines must meet will become evident if tie-lines are

* C. Wagner, Thermodynamics of the Liquidus and the Solidus of Binary Alloys, *Acta Metallurgica* **2** (2): 242–249 (1954).

drawn in the $L + \alpha$ region in the vicinity of the minimum point. Should the liquidus and solidus fail to meet, as in the impossible case depicted in Fig. 3-14a, some tie-lines in the $\alpha + L$ region would indicate equilibrium between solid and solid ($\alpha - \alpha$) instead of between liquid and solid. But if the liquidus and solidus meet at the minimum point (Fig. 3-14b), all the lines join conjugate liquid and solid compositions.

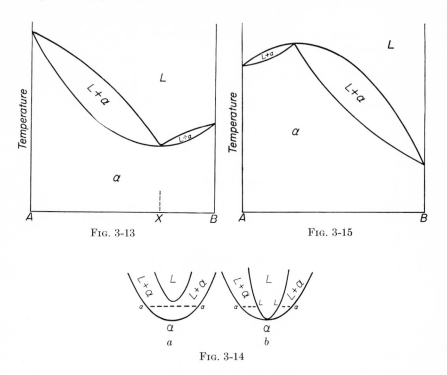

FIG. 3-13

FIG. 3-15

FIG. 3-14

The alloy which has the composition of the minimum melting point behaves much like a pure metal. It melts and freezes isothermally and exhibits no coring in its cast structure. Alloys such as this, which melt and freeze with the liquid and solid phases undergoing no changes in composition, are called *congruently melting alloys*. A congruently melting alloy may be treated as a component of an alloy system. Thus the system $AB$ in Fig. 3-13 may be divided into two systems at composition $X$. One will be the simple isomorphous binary system $AX$; the other the simple isomorphous binary system $XB$.

*Maxima* are also possible in isomorphous systems (see Fig. 3-15.) As in the case of the minimum, the liquidus and solidus meet tangentially at the maximum and the alloy of this composition is congruent in its melting and freezing behavior.

## PRACTICE PROBLEMS

**1.** Compare qualitatively the range of composition difference (coring) that would develop in the course of the natural freezing of a 90% Ni + 10% Cu alloy (Fig. 1-1) with that which would develop in a 50% Ni + 50% Cu alloy. What generalization can be deduced from this result?

**2.** Upon one set of coordinates, draw isomorphous diagrams for the two alloy systems *AB* and *AC*, where *A* melts at 1000°C, *B* melts at 800°C, and *C* melts at 200°C. Compare qualitatively the ranges of composition difference (coring) that would develop in the course of the natural freezing of an alloy from each system, where both alloys have the same content of component *A*. What generalization can be deduced from this result relating to the effect of the difference in the melting points of the component metals upon the extent of coring?

**3.** An isomorphous alloy system (*AB*) has a minimum at 70% *B*, as in Fig. 3-13 Predict the course of natural freezing of an alloy (*a*) at 70% B, (*b*) at 10% B, (*c*) at 95% B, (*d*) at 65% B.

# CHAPTER 4

# BINARY EUTECTIC SYSTEMS

In eutectic systems the first addition of either component metal to the other causes a lowering of the melting point, so that the liquidus curve passes through a temperature minimum known as the *eutectic point*. The liquid is miscible in all proportions, as in isomorphous systems, but miscibility in the solid state is limited (see Fig. 4-1).

Two solid phases, $\alpha$ and $\beta$, are to be distinguished. These are solid-solution phases, which are referred to as *limited solid solutions* because the range of stability of each extends only partway across the diagram and as *terminal solid solutions* because each originates at a side of the diagram and extends inward, instead of being isolated somewhere in the mid-part of the diagram. The $\alpha$ and $\beta$ phases *may* have identical crystal structures but usually do not; each, of course, has the crystal structure of the component with which it is associated.

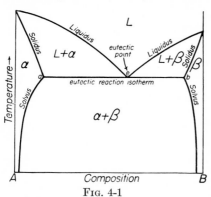

The phases are not restricted with regard to type of solid solution, i.e., whether substitutional or interstitial.

Three two-phase regions are designated in Fig. 4-1, namely, $L + \alpha$, $L + \beta$, and $\alpha + \beta$. It will be evident that the $L + \alpha$ and $L + \beta$ regions are in every way equivalent to the $L + \alpha$ region in the isomorphous diagram. These areas are considered to be made up of tie-lines joining the conjugate liquid and solid phases, represented on the liquidus and solidus, at each temperature within the range covered. In like manner, the $\alpha + \beta$ region is regarded as being made up of tie-lines connecting, at each temperature, a composition of $\alpha$, represented on the $\alpha$ *solvus* curve (or $\alpha$ solid-solution boundary) with its conjugate $\beta$ composition represented on the $\beta$ *solvus* curve (or $\beta$ solid-solution boundary).

The three two-phase regions meet at a special kind of tie-line that is common to all the regions and that joins the compositions of three conjugate phases that coexist at the eutectic temperature, namely, $\alpha$ at $a$, liquid at $e$, and $\beta$ at $b$. This tie-line is variously known as the *eutectic line*, the *eutectic horizontal*, and the *eutectic reaction isotherm*. Point $e$, representing the only liquid that can coexist simultaneously with both solid phases, is called the *eutectic point*, meaning the point, or composition, of lowest melting temperature.

### Application of the Phase Rule

The TX diagram of Fig. 4-1 is an isobaric section from a PTX space diagram, such as the one shown in Fig. 4-2, where the dashed section

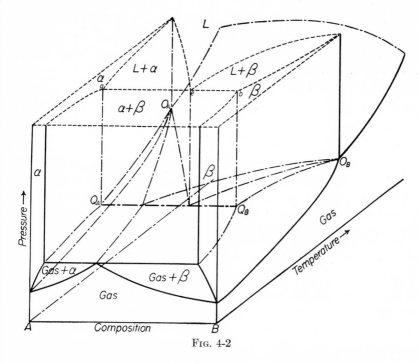

FIG. 4-2

across the top is the isobar represented in the previous figure. Again, one degree of freedom has been expended in establishing the pressure of the TX diagram. Hence, in the one-phase regions (liquid, $\alpha$ and $\beta$) which are tervariant, only the temperature and one composition remain to be selected in order to establish the equilibrium state. The two-phase regions are bivariant, so that a choice of temperature or of a composition, but not both, suffices to describe a fixed state of equilibrium. Both of these conditions have been discussed with reference to isomorphous equilibria and require no further comment here.

Three-phase equilibrium has been shown on the TX diagram at the eutectic reaction isotherm. Here univariant equilibrium obtains.

$$P + F = C + 2$$
$$3 + 1 = 2 + 2$$

Having used one degree of freedom in establishing the isobaric section, there should remain no further choice of conditions under which three phases may coexist. In other words, the three-phase equilibrium must be represented at a fixed temperature and the compositions of each of the three phases involved must also be fixed. The eutectic line satisfies these requirements; a horizontal line represents only one temperature on a TX diagram; the composition of the $\alpha$ phase is fixed at $a$, that of the liquid phase at $e$, and that of the $\beta$ phase at $b$. No one-phase region touches the eutectic line at more than a single point in composition.

There are four phases shown on the PTX diagram: $\alpha$, $\beta$, liquid, and gas. The phase rule permits a four-phase invariant equilibrium.

$$P + F = C + 2$$
$$4 + 0 = 2 + 2$$

Therefore, at some *one fixed condition* of pressure and temperature it is possible to have a four-phase equilibrium among $\alpha$, $\beta$, liquid, and gas, each of fixed composition. This occurs at the unique spatial tie-line $Q_A Q_B$ in Fig. 4-2. In most alloy systems this equilibrium must occur at very low pressure, but in any case, the likelihood of an arbitrarily selected isobaric section coinciding with the pressure of the four-phase equilibrium is so slight that it is usually considered that a binary TX diagram should not be expected to represent equilibrium among more than three phases.

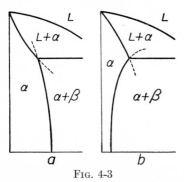

FIG. 4-3

## Some Other Rules of Construction

It can be shown through an application of the second law of thermo-dynamics (see Appendix V) that the boundaries of two-phase fields must meet at such angles that the extensions of the boundaries, beyond the point of intersection, project into two-phase fields, never into one-phase fields (see Fig. 4-3). Thus, the solidus and solvus curves must meet at the eutectic horizontal at such angles that the solidus projects into the $\alpha + \beta$ field and the solvus into the solid plus liquid field. This rule does not further restrict the direction of curvature of either line; i.e., the solidus may be either concave (Fig. 4-3$a$) or convex (Fig. 4-3$b$) with respect to the base of the diagram, and the solvus may be either concave (Fig. 4-3a) or convex (Fig. 4-3b) with respect to the side of the diagram. However, concavity of the solvus (Fig. 4-3b) seems to be most common in eutectic systems.

The shape of the solvus is frequently found to conform to the empirical relationship

$$\log \text{ at. } \% \ B = \frac{1}{T_k} + c$$

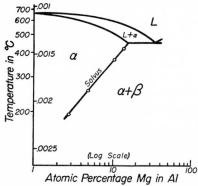

where $B$ is the solute metal, $T_k$ is the absolute temperature, and $c$ is a constant. If the solvus appears as a straight line when plotted with the logarithm of the atomic percentage as the abscissa and the reciprocal of the absolute temperature as the ordinate, it is reasonable to conclude that the data represent equilibrium conditions (see Fig. 4-4). Failure of the solvus to become a straight line in this test does not necessarily mean that it is unreliable.[1] It is generally found that the eutectic point lies closer to that component which has the lower melting point.

FIG. 4-4. Aluminum-rich portion of the aluminum-magnesium phase diagram; with log composition plotted versus reciprocal temperature (absolute), the solvus becomes straight.

## The Eutectic Alloy

When the eutectic alloy (composition $e$ in Fig. 4-1) is heated through the eutectic temperature, the $\alpha$ and $\beta$ phases (compositions $a$ and $b$) react to form liquid (composition $e$). Upon cooling through the eutectic temperature, the liquid decomposes into $\alpha$ and $\beta$.

$$L \underset{\text{heating}}{\overset{\text{cooling}}{\rightleftarrows}} \alpha + \beta$$

"Equilibrium" melting and freezing occur isothermally, as with a pure metal, but with the difference that the process is incongruent, which is to say that there is a difference in composition between the liquid and the individual solid phases.

During solidification the two solid phases are believed to deposit simultaneously. Barring the influence of undercooling, which is usually moder-

---

[1] For further details, see W. L. Fink and H. R. Freche, Correlation of Equilibrium Relations in Binary Aluminum Alloys of High Purity, *Trans. Am. Inst. Mining Met. Engrs.*, **111**:304–318 (1934).

ate in eutectic alloys, there is no coring effect, because freezing takes place at, or very close to, one temperature and the compositions of the phases are virtually constant. The relative proportions of the $\alpha$ and $\beta$ phases in the completely solidified alloy may be ascertained by treating the eutectic reaction isotherm (Fig. 4-1) as a tie-line.

$$\%\alpha = \frac{eb}{ab} \times 100 \approx 40\% \qquad \%\beta = \frac{ae}{ab} \times 100 \approx 60\%$$

With further cooling below the eutectic temperature, successive tie-lines in the $\alpha + \beta$ field call for changes in the compositions of the solid phases and also in their relative proportions. Such changes are rarely observed with normal rates of cooling, however, because of slow diffusion in

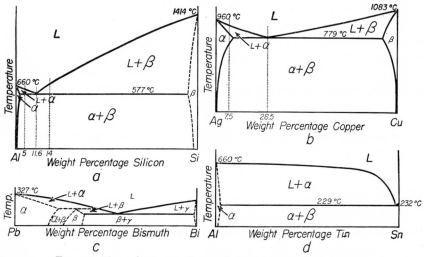

FIG. 4-5. Phase diagrams of some typical eutectic-type systems.

the solid state. The establishment of equilibrium at subeutectic temperature usually requires the use of heat treatments, which will be discussed presently.

The microstructure of eutectic alloys differs widely from system to system, depending, in part presumably, upon the relative proportions of the two solid phases and, in part, upon crystallographic relationships, which have yet to be elucidated. Among alloy systems having somewhere near equal proportions of the solid phases in the eutectic, Fig. 4-5b, for example, it often happens that the two phases are deposited in layers, or lamellae. Seen in cross section the lamellae appear as alternating bands of the $\alpha$ and $\beta$ phases (Fig. 4-6B). This is known as a pearlitic structure, because, as with a pearl, the structure may diffract light into an irregular pattern of iridescent colors.

When one of the phases predominates in the eutectic mixture (Fig. 4-5a, c, and d) this phase usually becomes a matrix in which particles of the lesser phase are embedded. Several types of eutectic structure are then distinguished upon the basis of the shapes assumed by the embedded

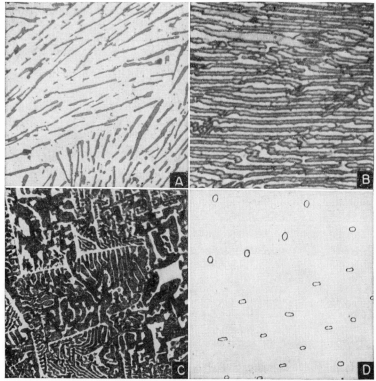

Fig. 4-6. Typical microstructures of some cast eutectic alloys: (A) Al-Si 11.6%, magnification 500; (B) Ag-Cu 28.5%, magnification 2,000; (C) Pb-Bi 56.5%, magnification 100; and (D) Al-Sn 99.5%, magnification 650. See also Fig. 4-5.

particles. If these particles are rounded, the eutectic is said to be of the nodular, globular, or spheroidal type (Fig. 4-6D). Elongated particles may appear nodular when seen in cross section and are sometimes included in this class. Flat platelets appear elongated, or needlelike, in cross section, and the eutectic structure formed by them is called acicular (Fig. 4-6A). An intricate pattern, known as *Chinese script* (Fig. 4-6C) results from the growth of the second phase in the form of obviously dendritic skeletons. Although differing in their structural appearance from system to system, the eutectics have the common characteristic that the compositions and proportions of the phases are constant for any one alloy system and the two solid phases form concurrently.[1]

## Hypoeutectic and Hypereutectic Alloys

Those alloys occurring upon one side of the eutectic composition are called hypoeutectic alloys, while those upon the other side are called hypereutectic alloys. There is no inflexible rule that can be used to determine, for any specific case, which side of the alloy system should be called hypoeutectic and which side hypereutectic. Custom favors the use of *hypoeutectic* for the designation of those alloys richer in the more common metal or the metal of more immediate importance in the application under consideration. Since most phase diagrams are drawn with the more common metal on the left, this means that the hypoeutectic alloys are usually those to the left of the eutectic composition while the hypereutectic alloys are those to the right.

Unlike the eutectic alloy itself, the hypo- and hypereutectic alloys undergo bivariant as well as univariant transformation during melting and freezing. The course of freezing passes through the zones of $L + \alpha$ or $L + \beta$ equilibrium, in addition to the eutectic $L + \alpha + \beta$ equilibrium. Hence, it will be desirable once more to distinguish between "equilibrium" and "natural" melting and freezing.

The course of equilibrium freezing may be followed by reference to Fig.

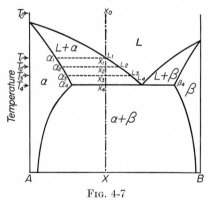

FIG. 4-7

4-7. At temperature $T_0$ the hypoeutectic alloy of composition $X$ is molten. Upon cooling the alloy to temperature $T_1$, freezing begins with the appearance of crystals having the composition $\alpha_1$. As freezing proceeds, with falling temperature, the composition of the liquid follows the liquidus to the right ($L_1$, $L_2$, $L_3$, $L_4$) and the composition of the solid, likewise, follows the solidus to the right ($\alpha_1$, $\alpha_2$, $\alpha_3$, $\alpha_4$), as in the equilibrium freezing of an isomorphous alloy. At the eutectic temperature $T_4$, however, this process is interrupted with some liquid remaining:

$$\%L_4 = \frac{\alpha_4 x_4}{\alpha_4 L_4} \times 100 \approx 50\%$$

[1] The development of eutectic structures is discussed in detail by F. L. Brady, The Structure of Eutectics, *J. Inst. Metals*, **28**:369–419 (1922); A. M. Portevin, The Structure of Eutectics, *J. Inst. Metals*, **29**:239–278 (1923).

The microstructure of such an alloy, at this moment, is depicted in the drawing of Fig. 4-8$A$, where the white areas represent the solid $\alpha$ and the black the remaining liquid. This liquid, which is of the eutectic composition, $L_4$, now freezes isothermally to a eutectic mixture of $\alpha + \beta$, Fig. 4-8$B$ (actually the 5% silicon alloy of Fig. 4-5$a$). The proportions of $\alpha$ and

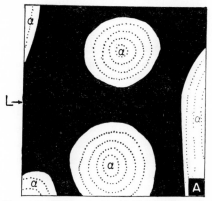 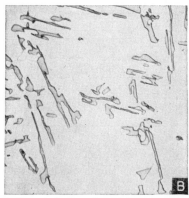

Fig. 4-8. Hypoeutectic Al-Si 5% alloy. Sketch $a$ indicates the location of the primary $\alpha$ and the eutectic liquid; the final microstructure at the end of freezing is shown in $b$. Magnification 500.

$\beta$ *in this mixture* are the same as in the eutectic alloy, (see, for example, Fig. 4-6$A$, the 11.6% silicon alloy of Fig. 4-5$a$), namely,

$$\%\alpha_4 \text{ (eutectic)} = \frac{L_4\beta_4}{\alpha_4\beta_4} \times 100 \approx 33\tfrac{1}{3}\%$$

$$\%\beta_4 \text{ (eutectic)} = \frac{\alpha_4 L_4}{\alpha_4\beta_4} \times 100 \approx 66\tfrac{2}{3}\%$$

But some $\alpha$ had formed during the initial stage of freezing, at temperatures above the eutectic. The *total* quantities of $\alpha$ and $\beta$ in the finally frozen alloy are, then,

$$\%\alpha_4 \text{ (total)} = \frac{x_4\beta_4}{\alpha_4\beta_4} \times 100 \approx 66\tfrac{2}{3}\%$$

$$\%\beta_4 \text{ (total)} = \frac{\alpha_4 x_4}{\alpha_4\beta_4} \times 100 \approx 33\tfrac{1}{3}\%$$

The $\alpha$ phase that appeared first is known as the *primary constituent* of the alloy; the $\alpha + \beta$ eutectic mixture is the *secondary constituent*. In its metallographic usage, the word constituent denotes a unit of the microstructure. It may be composed of a single phase, or of several phases, as in the case of the eutectic constituent. The adjectives primary and secondary refer to the order in which the constituents appear during freezing.

Failure to maintain equilibrium conditions in the *natural* freezing of the hypo- and hypereutectic alloys results in coring of the primary constituent and in an excess over the equilibrium quantity of the secondary con-

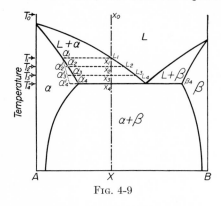

stituent. The reasons for these differences are similar to those discussed in the case of the natural freezing of isomorphous alloys and are illustrated in Fig. 4-9. During the primary separation of the $\alpha$ phase, the average composition of the solid follows a course of change such as $\alpha_1$, $\alpha_2'$, $\alpha_3'$, $\alpha_4'$, because diffusion in the solid state is not sufficiently rapid to maintain equilibrium except with almost infinitely slow cooling. Thus, the primary $\alpha$ phase is cored. At the eutectic temperature the quantity of liquid remaining is

Fig. 4-9

$$\%L_4 \text{ (natural)} = \frac{\alpha_4' x_4}{\alpha_4' L_4} \times 100 \approx 66\%$$

instead of

$$\%L_4 \text{ (equilibrium)} = \frac{\alpha_4 x_4}{\alpha_4 L_4} \times 100 \approx 50\%$$

Hence, there is a larger quantity of eutectic liquid than would be present in "equilibrium" freezing, and this produces a correspondingly larger quantity of the eutectic constituent. The photomicrograph of Fig. 4-8b, of course, represents a structure produced by "natural" freezing. Coring of the $\alpha$ phase, although really present and indicated by the dotted lines in Fig. 4-8a, is not evident in the photograph because there is no marked color change associated with composition difference in the aluminum-silicon solid solution. Neither is the line of demarcation between the primary and secondary constituents clearly defined in Fig. 4-8b. This is because the $\alpha$ phase predominates in the eutectic and because the last layer of $\alpha$ to form on the cored primary dendrites is of the same composition ($\alpha_4$) as the $\alpha$ phase of the eutectic constituent.

Hypereutectic freezing follows a similar course. Consider, for example, the aluminum-silicon alloy containing 14% silicon (Fig. 4-5a). Freezing begins with the primary separation of the $\beta$ (silicon-rich) phase. As the $\beta$ deposits, the liquid and solid compositions move toward the left until, at the eutectic temperature, the remaining liquid freezes isothermally to a secondary eutectic constituent. Two stages in this process are represented in Fig. 4-10. The primary $\beta$ crystals, which in this case are idiomorphic (exhibiting well-developed crystalline facets), are represented in black in the sketch, Fig. 4-10$A$, and appear gray in the photomicrograph, Fig. 4-10$B$. The eutectic liquid, white in the sketch, freezes to the now familiar eutectic mixture of $\alpha + \beta$. This time the line of demarcation between the

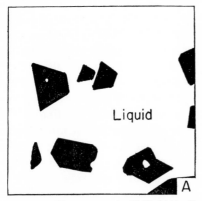
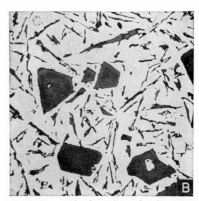

Fig. 4-10. Hypereutectic Al-Si 14% alloy. Sketch $A$ shows the location of primary $\beta$ and the eutectic liquid; photograph $B$ shows the final cast microstructure. Magnification 100.

primary $\beta$ constituent and the eutectic constituent is sharp, because the $\beta$ phase is not physically continuous in this structure and the $\alpha$ phase occurs only in the eutectic constituent.

## Terminal Solid Solutions

Under conditions of equilibrium freezing, alloys lying outside the composition range of the eutectic reaction isotherm, such as alloy $X$ of Fig.

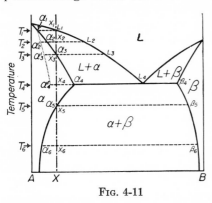

Fig. 4-11

4-11, should contain none of the eutectic constituent, because the process of freezing would be completed, at temperature $T_3$, before the composition of the liquid phase had reached that of the eutectic. Actually, however, such alloys, when solidified from the melt, do contain the eutectic constituent, except when the gross composition of the alloy lies *very close* to that of one of the pure metals. The explanation is, of course, to be found in the coring effect, which causes the average composition of the solid phase to depart so widely from the solidus that freezing is not completed before the eutectic temperature is reached. A small amount of liquid ($L_4$) remains:

$$\%L_4 = \frac{\alpha'_4 x_4}{\alpha'_4 L_4} \times 100 \approx 8\%$$

and this freezes isothermally as a secondary eutectic constituent of $\alpha + \beta$.

A typical alloy of this kind is sterling silver (7.5% copper, balance silver; see Fig. 4-5b), the cast microstructure of which is presented in

Fig. 4-12. Here the silver-rich $\alpha$ phase is light colored and the copper-rich $\beta$ is dark. Coring of the primary $\alpha$ dendrites is faintly visible in the shading of the light-colored areas, which are lightest at their silver-rich centers and become grayer toward their perimeters, where the copper content is higher.

In this example (Fig. 4-12) the secondary (eutectic) constituent, composed of $\alpha + \beta$, differs in appearance from the silver-copper eutectic (28.5% copper) shown in Fig. 4-6b in that its structure appears not to be pearlitic. Instead, somewhat irregular masses of the $\beta$ phase lie between the primary $\alpha$ dendrite arms. The portion of the $\alpha$ phase which belongs to the eutectic is physically continuous with the $\alpha$ of the primary constituent, so that it is difficult to discern where the boundary between the primary and secondary constituents lies. The eutectic structure is, in effect, composed of but one layer of the $\beta$ phase enclosed between two layers of the $\alpha$ phase which have grown directly upon the surfaces of the primary $\alpha$ masses. This is called a *divorced eutectic*. It is assumed that whereas the two solid phases of a normal eutectic are nucleated and grow together in a characteristic geometric pattern, one of the phases (the $\alpha$) of the divorced eutectic is nucleated by primary crystals of its own kind before new nuclei can form for eutectic growth, thereby altering the geometric arrangement of the phases.

FIG. 4-12. Sterling silver (7.5% Cu) as cast. Light areas, shading from white to light gray, are cored dendrite areas, silver-rich at their centers; dark areas are divorced eutectic. Magnification 100.

## Homogenization, or Solution Heat Treatment

If the terminal solid-solution alloy $X$ of Fig. 4-11 is brought to equilibrium by subjecting it to long heat treatment at a temperature between $T_3$ and $T_5$, it should, according to the phase diagram, be composed of the α phase alone (see Fig. 4-13). During this heat treatment the homogeneous condition is being approached in two ways simultaneously: (1) Coring of the α phase is being eliminated by a redistribution of the components, as in the homogenization of an isomorphous alloy, and (2) the β phase is dissolving in the α phase.[1] The rate at which these changes proceed will depend, of course, upon the temperature of heat treatment, because both changes are effected by diffusion and the diffusion velocity doubles, approximately, in most alloys for each temperature rise of 50°C. By quickly cooling the homogenized alloy to room temperature it is usually possible to retain the homogeneous condition, at least temporarily. The phase diagram indicates that the β phase should again be stable in this alloy at room temperature, but except in metals of low melting point, the rate of

Fig. 4-13. Same as Fig. 4-12, homogenized by heat treatment just below the eutectic temperature and quenched to room temperature; only the silver-rich solid solution is present in this structure. Magnification 100.

diffusion at room temperature is ordinarily so slow that the second solid phase does not reappear for a very long time.

The same heat treatment applied to an alloy within the span of the eutectic reaction isotherm, such as alloy $X$ in Fig. 4-9, will not cause the β phase to disappear, but coring of the α phase can be eliminated, and any excess over the equilibrium quantity of the β phase will dissolve in the α. In addition, the remaining particles of the β phase may become *spheroidized*, i.e., assume rounded shapes of minimum surface.

### Precipitation in the Solid State

As has been remarked above, the homogenized terminal solid solution when cooled below $T_5$ (Fig. 4-11) becomes unstable and tends to reject the β phase. Actual precipitation of the β phase takes place if the temperature is high enough so that the rate of diffusion is appreciable. This condition may be realized either by slow cooling from high temperature or by re-heat-treating at a temperature only a little below $T_5$. The β precipitate usually takes the form of small idiomorphic particles situated along the grain boundaries and within the grains of the α phase (Fig. 4-14). In the majority of cases it is evident, also, that the β particles,

[1] If the temperature is above $T_4$, the β phase will, of course, be replaced by liquid and homogenization will proceed by the solution of liquid in the α phase.

having a more or less regular form and size, are arranged in a systematic fashion with respect to orientation. Many of the particles within a single

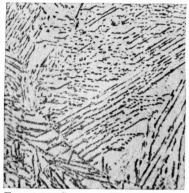

FIG. 4-14. Same as Fig. 4-13, but slowly cooled from the homogenizing temperature. Parallel platelets of copper-rich solid solution have formed upon certain crystallographic planes of the parent silver-rich solid solution, producing a Widmanstätten structure. Magnification 1,000.

α grain lie parallel to one another, and if the precipitate particles be assigned to groups according to the direction assumed, it is found that there is a limited number of such groups (1, 3, 4, 6, 12, 24). This is called a *Widmanstätten structure.* The newly formed crystals of the β phase take their orientations from the parent α crystal within which they grow.[1] In the sketch of Fig. 4-15a the particles are drawn as though they were square platelets placed parallel to the cube planes of the parent crystal, in the manner indicated by Fig. 4-15b. Since the parent α crystals occur in various orientations in the microstructure, the traces of the β platelets likewise assume different orientations from one α grain to the next, but the number of directions in any one α grain is always the same (three in this example).

Precipitation in the solid state occurs by the formation of oriented nuclei which grow at a rate depending upon the speed with which the appropriate variety of atoms is delivered by diffusion. Therefore, the rate of precipitation tends to be very slow at room temperature and to increase as the temperature is raised. The compositions of the participating phases also change somewhat with temperature as can be seen by reference to Fig. 4-11, where the solvus curves slope inward as the eutectic temperature is approached. At any stated temperature, such as $T_6$, the compositions and relative proportions of the α and β phases can be computed from the tie-line connecting the two solvus curves.

The precipitation of a second phase from a solid solution is sometimes accompanied by a marked change in mechanical properties. In general, the hardness and strength decrease during precipitation because the solid solution is becoming more dilute. But in the *age-hardening alloys,* there is a substantial increase in hardness and allied properties. This increase is associated with the formation of a transition state that forms first. Subsequently this may change into the stable precipitate, with a reversal in the direction of the property change. The transition state, being unstable

[1] For further details of the Widmanstätten structure, see C. S. Barrett, "Structure of Metals," chap. 22, pp. 538–580, McGraw-Hill Book Company, Inc., New York, 1952.

at all temperatures and compositions, is not distinguished upon the phase diagram, and there is no known thermodynamic criterion by which

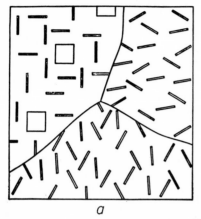
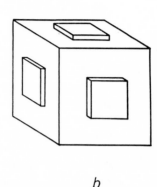

FIG. 4-15. Schematic representation of precipitate particle arrangement in a Widmanstätten structure. A matrix crystal with platelets parallel to three cube faces is represented in *b*; the appearance of these platelets in a microstructure including three differently oriented matrix crystals is shown in *a*.

its existence could be predicted. From the viewpoint of alloy constitution it can be said only that age hardening becomes possible when the solid solubility decreases with falling temperature, but among the many alloys having this characteristic only a limited number exhibit appreciable age hardening.

## Liquation

Partial melting, which may occur inadvertently in the course of heat treatment and which is known as liquation, or burning, of the alloy, commonly proceeds with the forma-

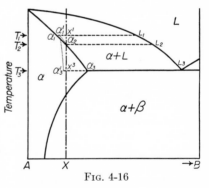

FIG. 4-16

tion of a continuous film[1] of liquid along the grain boundaries of the $\alpha$ phase and sometimes also in pockets between the dendrite arms. If the terminal solid-solution alloy $X$ in Fig. 4-16 has previously been homogenized, the first liquid to form upon liquation will have the composition $L_2$. As the temperature rises above the solidus, the liquid becomes richer in the $A$ component; the $\alpha$ phase also becomes enriched in the $A$ component, but its average composition lags

---

[1] Continuous film formation occurs when the liquid phase "wets" the solid phase. This condition appears to be very common among metal systems.

behind the increase demanded by the solidus. Thus the percentage of the liquid phase is less than the equilibrium quantity at any given temperature such as $T_1$:

$$\%L_1 = \frac{\alpha_1' x_1}{\alpha_1' L_1} \times 100 \approx 10\%$$

If the alloy is now cooled, the liquid composition proceeds in the reverse direction down the liquidus. Cored $\alpha$ phase is deposited until the eutectic temperature is reached, where any remaining liquid is converted to the eutectic constituent $\alpha + \beta$. In this way liquation may cause the reappearance of the eutectic constituent, and the work of the solution heat treatment is thereby undone.

Because the liquid existed as a continuous grain boundary film during liquation and because the film was relatively thin, the resulting eutectic constituent is usually highly divorced. Consequently, liquation may result in the formation of an almost *continuous grain boundary network* of the $\beta$ phase, Fig. 4-17. Any liquid existing in interdendritic pockets freezes in place as eutectic *rosettes*.

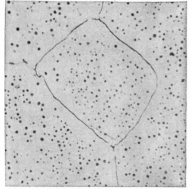

FIG. 4-17. Same as Fig. 4-13 reheated to slightly above the solidus temperature of the alloy and quickly cooled. Liquation at grain boundaries and at many isolated points within the grains has resulted in the development of a continuous film of eutectic copper solid solution at the grain boundaries and eutectic rosettes within the grains. Magnification 100.

When overheating occurs, before homogenization has progressed to the disappearance of the $\beta$ phase (or when dealing with an alloy within the composition span of the eutectic isotherm), melting begins at the eutectic temperature and the first liquid is of the eutectic composition. For this reason, the temperature of the homogenizing heat treatment is usually established below the eutectic temperature. After the $\beta$ phase is dissolved, the temperature may be raised gradually to just below the solidus without the advent of liquation. By rehomogenizing, the segregation caused by liquation can usually be eliminated.

The time required is often much longer, however, than is needed for the homogenization of the original cast structure. This is because the bulk of the diffusion required to redistribute the $B$ component must operate over the relatively long path from the grain boundary to the interior of the grain whereas in the "as-cast" structure the major part of the $\beta$ phase resides in interdendritic spaces and the diffusion path is accordingly shorter.

## "Modification" of the Eutectic

There are several eutectic alloys that are subject to a marked lowering of the apparent eutectic temperature either when cooled very quickly from the molten condition or when cooled at a normal rate after the addition of some special reagent. Not only is the eutectic temperature lowered, but the composition of the eutectic point is shifted and the particle size of the eutectic constituent is refined. This effect is known as the "modification" of the eutectic.

Aluminum-silicon alloys are among the best known of those that are subject to modification. If a small quantity of sodium is added to the

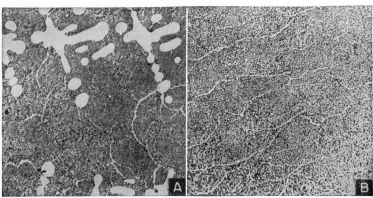

FIG. 4-18. (A) Aluminum-silicon eutectic alloy (11.6% silicon, see Fig. 4-6A) modified with sodium. The structure is now hypoeutectic; light areas are primary aluminum-rich solid solution, dark areas the modified eutectic constituent. Magnification 100. (B) Aluminum-silicon hypereutectic alloy (14% silicon, see Fig. 4-10B) modified with sodium. The structure is now composed altogether of the modified eutectic constituent. Magnification 100. (*Courtesy of The Aluminum Company of America.*)

aluminum-silicon eutectic alloy (11.6% silicon, see Fig. 4-5a), or if the eutectic alloy is rapidly chilled from the melt, the cast microstructure will appear somewhat as in Fig. 4-18A, which should be compared with the photograph of the normal cast structure of this alloy as shown in Fig. 4-6A. It is apparent from the microstructure that this alloy has become hypoeutectic; there are large primary dendrites of the $\alpha$ phase, which were not present in the unmodified material. The alloy containing 14% silicon was seen to be hypereutectic in its normal cast state, Fig. 4-10B, but when modified, it appears as though of eutectic composition, Fig. 4-18B. Modification has shifted the eutectic point from 11.6 to 14% silicon (Fig. 4-19). At the same time the eutectic temperature has been reduced. The extent of its reduction depends upon the conditions of modification, the lowest value regularly observed being about 555°C.

If the aluminum-rich liquidus is projected, it is found that it passes very close to 14% silicon at 555°C. From this it has been concluded by some that modification results from the undercooling of the β phase, so that bivariant crystallization of the α phase can continue below the normal eutectic temperature and until the β phase begins to appear at some

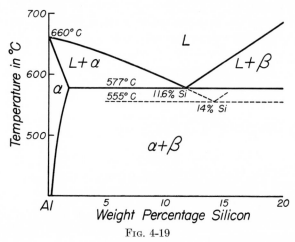

Fig. 4-19

lower temperature. The rate of nucleation being greater at lower temperature, a finer eutectic structure is produced.[1]

## Isothermal Diffusion Structures

Progress from nonequilibrium toward equilibrium at constant temperature is well illustrated by diffusion couples. Consider, for example, a piece of pure copper held at some elevated temperature until a stable state is established. If now a piece of silver is laid upon the copper (with good contact), equilibrium will have been destroyed, because pure copper and pure silver cannot coexist at equilibrium. Silver will begin to dissolve in the copper, and the copper will begin to dissolve in the silver. Dissolution will progress until one of the metals has wholly dissolved in the other or until both become saturated at the temperature concerned. Similar processes are met in welding and soldering or when a structure composed of two or more metals is heated or when metal powders are mixed and heated to sinter them together.

[1] For details, see R. S. Archer and L. W. Kempf, Modification and Properties of Sand Cast Aluminum-Silicon Alloys, *Trans. Am. Inst. Mining Met. Engrs.*, **73**:587 (1926); B. Otani, Silumin and Its Structure, *J. Inst. Metals*, **36**:243 (1926); A. G. C. Gwyer and W. H. L. Phillips, The Constitution and Structure of the Commercial Aluminum Alloys, *J. Inst. Metals*, **36**:283 (1926); L. Guillet, Aluminum-Silicon Alloys and Their Industrial Uses, *Rev. mét.*, **19**:303 (1922).

The application of the phase diagram to predict alloy behavior in such cases is demonstrated in Fig. 4-20. Two examples are shown. At a temperature of 700°C (below the eutectic) copper and silver are brought together. The two metals become mutually saturated at their interface, and if the time for diffusion is short, the composition in each metal will vary from saturation at the interface to pure metal at locations remote from the interface. These are the composition ranges shown by the heavy lines *gh* and *ij*. No two-phase structure corresponding to the $\alpha + \beta$ field is produced; instead, the interface where the $\alpha$ and $\beta$ phases meet represents the

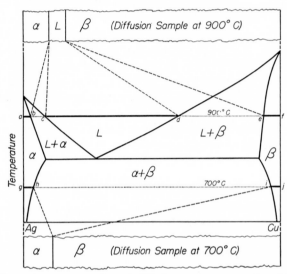

Fig. 4-20. Schematic representation of diffusion structures produced by isothermal diffusion between silver and copper at two temperatures. The top bar represents the layers produced at 900°C, the dashed lines associating these layers with the corresponding composition ranges upon the phase diagram. The bottom bar gives the same information for a diffusion temperature of 700°C.

entire two-phase region. This must be so because the only compositions of $\alpha$ and $\beta$ which may coexist at this temperature are those at *h* and *i*, and these compositions exist in the metal only at the interface. From this observation a general principle emerges, namely, that *isothermal diffusion between two metals results in the formation of one-phase layers corresponding to all one-phase fields crossed by drawing a line across the phase diagram at the temperature of diffusion, and the layers occur in the same sequence as the fields on the phase diagram.*

At 900°C (above the eutectic) three one-phase fields are crossed: $\alpha$, liquid, and $\beta$. The corresponding structure will be composed of a layer of the silver solid solution separated from the $\beta$ by a layer of liquid. In each layer the composition varies across the one-phase field, as represented on the phase diagram, namely, $\alpha$ from $a$ to $b$, liquid from $c$ to $d$, and $\beta$ from $e$ to $f$. Thus, if a piece of silver-plated copper is heated above the eutectic temperature, melting will occur at the junction between the two metals; this is far below the melting point of either pure metal.

If diffusion is allowed to proceed for a longer time at either temperature considered above, copper will eventually diffuse to the far side of the silver and silver to the far side of the copper. The range of compositions appearing in the sample will then extend from somewhere inside the $\alpha$ field to somewhere inside the $\beta$ field. If the quantity of silver is small compared with that of the copper, at the lower temperature, the silver will be consumed entirely in saturating the copper and only one phase, the $\beta$, will remain. At the higher temperature, if the gross composition of the composite sample lay between $e$ and $f$, the sample would, in time, be converted wholly to $\beta$; if it lay between $c$ and $d$, the entire sample would eventually melt; or if between $d$ and $e$, equilibrium would be reached when saturated copper of composition $e$ was in contact with liquid of composition $d$.

In practice it sometimes happens that diffusion structures are produced while the temperature is changing. Under such conditions the structure that is developed may be understood by combining the principles of isothermal phase change with those of phase change accompanying temperature variation. For example, if the material represented in Fig. 4-20 at 900°C is cooled to room temperature, the liquid layer must freeze to a series of alloys having compositions ranging from $c$ to $d$, thereby forming a two-phase layer of $\alpha + \beta$ between the one-phase zones of $\alpha$ and $\beta$. This is, obviously, *not an exception* to the rule that two-phase layers are never formed by isothermal diffusion in a binary system.

### Physical Properties of Eutectic Alloys

The properties of alloys of the eutectiferous systems depend both upon the individual characteristics of the phases and upon the mode of distribution of these phases in the microstructure. This is true, of course, of any multiphase alloy. Individually the phases behave as solid solutions whose properties vary within the composition range covered in the same manner as do those of the isomorphous solid solutions. That is, the property change is most rapid with the first addition of the solute to the solvent component, each successive addition augmenting the change to a diminishing degree. Hardness and tensile strength increase with alloying; elongation and electrical conductivity decrease.

With two phases coexisting, the composition, and hence the physical properties of the individual phases, "ideally" should remain constant across the entire range of two-phase equilibrium. The properties of the alloy should be simply those of a mixture of two substances of fixed characteristics. Departure from equilibrium interferes with the realization of this condition in many cases, but there are also many cases where the condition is realized or very nearly so. Obviously, there may be, and usually are, great differences between the properties of the two solid phases. If a phase is an intermediate solid solution (to be discussed in a later chapter), it is often very hard or brittle or both. Terminal solid solutions of the more common metals are usually relatively soft and ductile.

When two phases occur together in the structure, the resulting properties of the mixture most nearly resemble those of the physically continuous phase, i.e., that phase which forms the matrix in which particles of the other phase are embedded. As the quantity of the embedded phase increases with composition change, it tends toward a state of physical continuity and its properties are gradually approached. Thus the properties change more or less regularly with composition in the two-phase alloy from those of one of the saturated phases to those of the other. The change is almost linear in some systems (Fig. 4-21), but in others it deviates widely from this course.

Such deviations may be assignable to one or more of several causes, including (1) inhomogeneity of the individual phases, causing their properties to continue to change within the composition range of two-phase equilibrium; (2) peculiarities of particle shape that cause one of the phases to remain physically continuous over a disproportionately large range of composition; and (3) variations in particle size across the two-phase zone. The latter effect has two aspects, the size of the individual particles of the phases and the grain size within each particle. In cast alloys the individual particles are most commonly single crystals, but mechanical and thermal treatments, subsequently applied, may cause them to recrystallize to polycrystalline particles. A fine particle size is to be associated with greater hardness and decreased ductility but has little influence upon other than the mechanical properties. It is frequently observed that a maximum or minimum occurs in a property at the eutectic composition. This is presumed to be the result of a minimum particle size that is often found in eutectic and near-eutectic alloys.

Large effects of heat treatment upon the physical properties can, as a rule, be associated with corresponding changes in microstructure. During the homogenization of a terminal solid-solution alloy the quantity of dissolved solute is increased and the solid solution is thereby hardened. The reverse process takes place during precipitation unless age hardening intervenes. Spheroidization increases the physical continuity of the matrix phase and increases its influence upon the mechanical properties. Changes in the grain or particle size tend to harden or soften the alloy accordingly as the change is toward finer or coarser size. If liquation during heat treatment causes the lesser phase to be distributed in a film at the grain boundaries of the major phase, then the properties of the alloy will become predominantly those of the lesser phase, even though it be present in very small proportion. And if, in addition, the lesser phase is brittle, the alloy will be brittle—hence the expression "spoiled by burning."

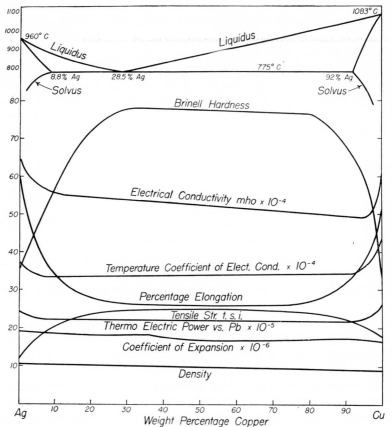

Fig. 4-21. Properties of annealed silver-copper alloys at room temperature. (*From Broniewski and Weslowski.*)

These principles of correlation between microstructure and physical properties apply alike to all multiphased alloys. Differences that are observed among typical alloys of the several constitutional types discussed in this and succeeding chapters are generally to be associated with differences in the relative importance of the several structural factors that have been mentioned. Eutectic alloys only occasionally exhibit large property effects that can be associated with the particle size as a major influence, while the eutectoid alloys about to be considered exhibit very important particle-size effects. The subject having been discussed in some detail in this and the preceding chapter, consideration of the correlation of the physical properties with constitution and microstructure will be minimized in succeeding chapters.

## Limiting Cases of the Eutectic

There are many alloy systems in which the eutectic composition occurs very close to one of the components (Fig. 4-22, see also Fig. 4-5d). The eutectic constituent is then composed predominantly of one phase in which a few widely separated particles of the second phase are embedded. Such a eutectic alloy closely resembles the dominant pure metal in its properties.

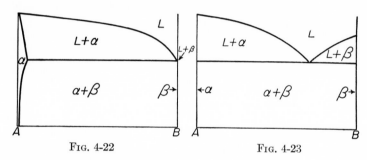

FIG. 4-22                    FIG. 4-23

The range of terminal solid solution is sometimes very narrow, so narrow, in fact, that it cannot be represented on the phase diagram because its width is less than that of the boundary line used to enclose it. In such cases, the eutectic reaction isotherm is extended to the composition of the pure component (Fig. 4-23). It should not be assumed, however, that this means that there is no range of solid solubility, for it is inconceivable that any two elements should be totally insoluble in one another at temperatures above absolute zero. Where the solid-solubility range is very small, coring in the primary constituent becomes minute because no substantial range of solid solutions can exist, but the likelihood of a eutectic constituent appearing in the microstructure is increased by low solid solubility.

### PRACTICE PROBLEMS

**1.** In the eutectic alloy system $AB$ the compositions of the three conjugate phases of the eutectic are $\alpha = 15\% B$, $L = 75\% B$, $\beta = 95\% B$. Assuming equilibrium freezing of an alloy composed of equal parts of $A$ and $B$ to a temperature infinitesimally below that of the eutectic, compute (a) the percentages of primary $\alpha$ and eutectic $\alpha + \beta$, (b) the percentages of total $\alpha$ and total $\beta$. How would each of the above answers be affected if equilibrium were not maintained during freezing?

**2.** Draw a eutectic diagram for the following case: $A$ melts at 1000°C, $B$ melts at 700°C, an alloy composed of 25% $B$ is just completely frozen at 500°C and at equilibrium is made up structurally of $73\frac{1}{3}$% of primary $\alpha$ and $26\frac{2}{3}$% of eutectic $\alpha + \beta$, whereas an alloy composed of 50% $B$ at the same temperature is made up structurally of 40% of primary $\alpha$ and 60% of eutectic $\alpha + \beta$, the total of $\alpha$ in this alloy being 50%.

**3.** A certain eutectic alloy system has a completely symmetrical phase diagram (that is, $A$ and $B$ have the same melting temperature, the eutectic point is at 50% $B$, and the solubility limits of the terminal solid solutions are identical). By means of long heat treatment, an alloy may be brought to equilibrium at a series of temperatures below the eutectic. When this is done with the eutectic alloy (50% $B$), no change is observed in the relative quantities of $\alpha$ and $\beta$ in going from one temperature to another, but an alloy composed of 25% $B$ exhibits a progressive change in the proportion of the two phases. Why should this be so? What change, if any, is occurring in the eutectic alloy?

CHAPTER 5

# BINARY EUTECTOID SYSTEMS

By substituting other phases for the liquid and two solids of the eutectic reaction, related types of systems are obtained which differ only in the nature of the phases involved. One of the most important of these is the eutectoid reaction, involving three solid phases (Fig. 5-1). Upon cooling, one solid phase ($\gamma$) decomposes into two other solid phases ($\alpha$ and $\beta$), and with heating, the reverse of this reaction occurs.

$$\gamma \underset{\text{heating}}{\overset{\text{cooling}}{\rightleftarrows}} \alpha + \beta$$

This type of system can be shown to conform with the requirements of the phase rule by applying the same arguments as those described in the case of eutectic systems. The form and directions of the phase boundaries

which terminate upon the eutectoid reaction isotherm are also governed by the same consideration as those cited for the phase boundaries terminating upon the eutectic reaction isotherm. It should be noted that in the example of Fig. 5-1, both of the component metals undergo allotropic transformation; the $A$ component transforms from $\alpha$ at low temperatures to $\gamma$ at high temperatures, and the $B$ component from $\beta$ to $\gamma$; the $\gamma$'s are isomorphous. Other examples in which the eutectoid reaction occurs without

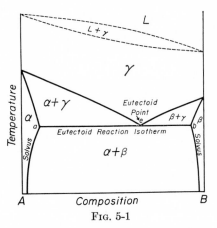

Fig. 5-1

the presence of allotropic transformations will be mentioned later. Those portions of the diagram of Fig. 5-1 which are drawn with dashed lines are not involved in the discussion of the eutectoid. A liquidus and solidus are included merely to complete the diagram, but there are several other constructions that could as well have been used for this purpose.

57

No generally applicable names have been applied to the boundaries of the $\alpha + \gamma$ and $\beta + \gamma$ fields. In the iron-carbon system, Fig. 5-2, which includes a eutectoid reaction, the $\gamma$, or upper, boundary of the $\alpha + \gamma$ field is commonly called the $A_3$ line, and the $\gamma$ boundary of the $Fe_3C + \gamma$ field (where $Fe_3C$ corresponds to $\beta$) is called the $A_{cm}$ line. No names at all have been given to the lower boundaries of these fields. In subsequent discussions these boundaries will be referred to as the $\gamma$ (upper) and $\alpha$ (lower) boundaries of the $\alpha + \gamma$ field and the $\gamma$ (upper) and $\beta$ (lower) boundaries of the $\beta + \gamma$ field. The lateral boundaries of the $\alpha + \beta$ field are properly termed *solvus* curves, as in the eutectic systems. The *eutectoid reaction isotherm* is also known by the names *eutectoid line* and *eutectoid reaction horizontal* and, in the iron-carbon alloys, as $A_1$.

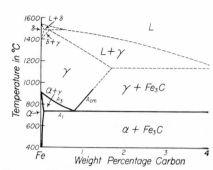

FIG. 5-2. Phase diagram of the metastable system iron-carbon.

## Structure of the Eutectoid Alloy

Transformations involving the dissociation and reassociation of solid phases are much more sluggish than similar transformations involving one or more liquid phases because diffusion rates are much slower in the solid state. Nonequilibrium transformation assumes a correspondingly greater practical significance and must be considered in greater detail than was found necessary with eutectic alloys. In order to provide a logical starting point for this discussion, the structures produced by equilibrium transformation, corresponding to infinitely slow reaction, will be dealt with first.

The *eutectoid alloy*, composition $e$ in Fig. 5-1, would undergo isothermal incongruent transformation in the "ideal" case. In every eutectoid system which has been studied it has been found that the high-temperature phase $\gamma$ decomposes upon cooling to produce a lamellar, or "pearlitic," eutectoid constituent composed of the $\alpha$ and $\beta$ phases (Fig. 5-3B). There is little to differentiate the equilibrium eutectoid and eutectic pearlitic structures in so far as appearance is concerned. Both form in so-called *colonies*, within which the lamellae are approximately parallel, and the crystallographic orientation of each of the two phases concerned is the same. In cast alloys the colonies tend to become elongated in the direction of freezing, while in the eutectoid structures the colonies are more commonly equiaxed (equal length and breadth), but there is insufficient regularity of behavior in this respect to make this difference a useful distinguishing criterion.

## Hypoeutectoid and Hypereutectoid Alloys

The terminology applied to alloys on the right and left of the eutectoid composition parallels that used for the corresponding alloys of eutectic systems. Those alloys lying to the left of the eutectoid point, or on the side of the more important metal, are designated *hypoeutectoid;* those on the opposite side are *hypereutectoid.* Transformation in these alloys proceeds in two steps (Fig. 5-4): First, a proeutectoid precipitation of one of the low-temperature phases occurs over a temperature range, followed by an "isothermal" decomposition of any remaining portion of the high-temperature phase into the eutectoid constituent (Fig. 5-3$A$ and $C$). Here the term *proeutectoid* is used with reference to the tranformation among solid phases in the same sense as is "primary" with respect to the freezing process.

If the hypoeutectoid alloy $X$, Fig. 5-4, is heated to $T_0$ and is held until equilibrium is obtained, it will be composed solely of crystals of the $\gamma$ phase. Lowering the temperature to $T_1$ should cause an incipient precipitation of $\alpha$ crystals of composition $\alpha_1$. Since the nucleation of precipitating phases is usually stimulated by surfaces, including grain boundaries, it is to be expected that there will be a marked tendency for the first crystallites of $\alpha_1$

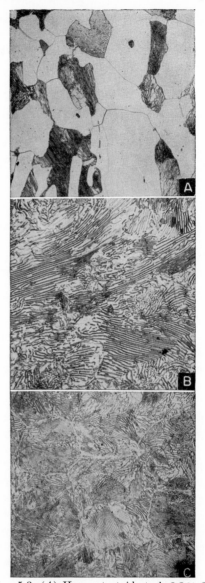

FIG. 5-3. ($A$) Hypoeutectoid steel, 0.3% C. Light areas are proeutectoid ferrite ($\alpha$); dark areas are the pearlitic eutectoid constituent ($\alpha + Fe_3C$). Magnification 500. ($B$) Eutectoid steel, 0.8% carbon, typical pearlite ($\alpha + Fe_3C$). Magnification 500. ($C$) Hypereutectoid steel, 1.2% C. Thin bands of proeutectoid cementite ($Fe_3C$), light gray, outline the grains of pearlitic eutectoid constituent ($\alpha + Fe_3C$). Magnification 500.

to appear at the grain boundaries of the parent $\gamma$ structure, and this is observed. As cooling proceeds, the phase diagram demands that the $\alpha$ crystals grow and at the same time change in composition along the $\alpha$ boundary of the $\alpha + \gamma$ field until the composition $\alpha_3$ is reached at the

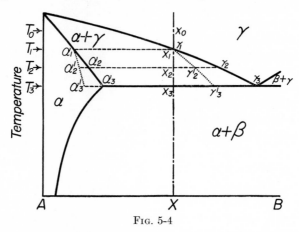

FIG. 5-4

eutectoid temperature $T_3$. Simultaneously, the $\gamma$ phase, if it were to respond fully to the requirements of equilibrium, would change in composition along the $\gamma$ boundary of the $\alpha + \gamma$ field until it attained eutectoid composition $\gamma_3$. All of the remaining $\gamma$ phase

$$\%\gamma_3 \text{ (equilibrium)} = \frac{\alpha_3 x_3}{\alpha_3 \gamma_3} \times 100 \approx 45\%$$

should now be converted to the eutectoid constituent.

The resulting microstructure as exemplified by an alloy of iron with 0.30% carbon (equivalent to about 5% $Fe_3C$) appears in Fig. 5-3A. It will be noted that the eutectoid pearlite occurs in separate "patches" or "nodules," each composed of one or more "colonies." The patches of pearlite are surrounded by the white $\alpha$ phase, which is nearly continuous in the microstructure because it grew preferentially from nuclear sites upon the grain boundaries of the $\gamma$ phase, which are now entirely gone.

A similar course of transformation in a hypereutectoid steel containing 1.20% carbon (equivalent to 18% $Fe_3C$) leads to the proeutectoid separation of $Fe_3C$ (the $\beta$ phase of Fig. 5-1). Because there is relatively much less of the proeutectoid constituent at this gross composition, it remains as a thin white network outlining the original $\gamma$ crystals in Fig. 5-3C. All remaining $\gamma$ has been transformed to eutectoid pearlite, as in previously considered cases.

By analogy with the natural freezing of hypoeutectoid alloys, it should be expected that in the transformation of a hypoeutectic alloy, the $\alpha$

phase will fail to maintain its equilibrium composition and that its average composition will follow a path such as $\alpha_1\alpha_2'\alpha_3'$ in Fig. 5-4. The differences in compositions involved are so small in all the systems that have been investigated, however, that no information is available on the structural nature of the hypothetical "eutectoid coring." The $\gamma$ phase, being crystalline, should also be expected to depart from its equilibrium composition during transformation, and a path such as $\gamma_1\gamma_2'\gamma_3'$ may be followed.

## Transformations at Fixed Subcritical Temperature

Slow reaction in the solid state usually makes it possible to cool the eutectoid alloy to temperatures far below the eutectoid temperature without decomposition of the high-temperature phase, provided, of course, that the speed of cooling is sufficient. If the alloy is quickly cooled to and then held at a temperature somewhat below that of the eutectoid, transformation will proceed at a definite rate characteristic of that temperature. This may be done by removing the piece from the furnace in which it has been previously converted to the $\gamma$ phase and immediately immersing it in a constant-temperature bath, such as a lead, salt, or oil bath, maintained at a controlled temperature. As the temperature of this kind of subcritical (below eutectoid) isothermal transformation is further and further decreased below the eutectoid temperature, the lamellae of the pearlite that is formed become thinner and closer together and the rate at

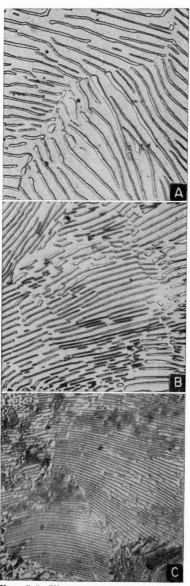

Fig. 5-5. Illustrates the variation in pearlite lamella spacing with temperature of formation: $(A)$ formed at the highest temperature, $(B)$ formed at an intermediate temperature, and $(C)$ formed at the lowest temperature. Magnification 1,500.

which the transformation proceeds becomes more rapid. This change in the "fineness" of the pearlite is illustrated, for a eutectoid steel, by the series of photographs in Fig. 5-5, and the variation in the rate of transformation with temperature of this steel is recorded in Fig. 5-6. In the latter diagram, temperature is plotted vertically, time of transformation

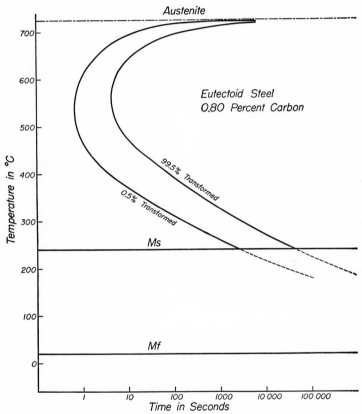

Fig. 5-6. Typical time-temperature-transformation (TTT) curve.

horizontally. The left-hand curve of the pair plotted represents the beginning of transformation (i.e., 0.5% pearlite); the right-hand curve the completion of transformation (i.e., 99.5% pearlite). In simple carbon steels, pearlitic structures result from isothermal subcritical transformation at all temperatures down to the first reversal (knee) of the curves at about 550°C. Pearlite nuclei form preferentially at the grain boundaries of the $\gamma$ crystals, and the rate of isothermal transformation is, accordingly, the greater the finer the austenite ($\gamma$) grain size.

Below the knee of the reaction curves, shown in Fig. 5-6, a different mechanism of transformation appears. The $\gamma$ phase decomposes into

$\alpha + \beta$, but the resulting structure is not lamellar. Corresponding to the colonies of pearlite are acicular (needle-shaped) zones made up of $\alpha + \beta$ in a state of unresolvably fine distribution (Fig. 5-7). This structure in steels is called *bainite*. That the rate of this type of transformation decreases with falling temperature is shown by the curves of Fig. 5-6.

Finally at low temperature, the reaction curves are interrupted by a pair of horizontal lines ($M_s$ and $M_f$) which correspond to the suppression of the $\gamma \rightarrow \alpha + \beta$ reaction and the appearance, in its stead, of a transformation of $\gamma$ to an unstable transition phase, called *martensite* (Fig. 5-8). There is no "rate of transformation" of $\gamma$ to martensite, no diffusion

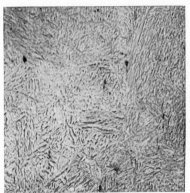

FIG. 5-7. Bainite constituent in a partially transformed carbon steel. Magnification 1,500.

FIG. 5-8. Martensite constituent in a quenched low-carbon steel. Magnification 500.

being involved in this transformation. The reaction curves are horizontal; only with lowering temperature does the amount of martensite increase from zero at the "beginning line" $M_s$ to 100% at the "ending line" $M_f$. At lower temperatures no further structural change is observed.

The entire graph of Fig. 5-6 is known as an *S curve* or, more properly, the TTT (time-temperature-transformation) curve. All the eutectoid alloys that have been studied have been found to transform in the same way, though with complications in some instances, and hence, a TTT curve may be used to describe the rate of each eutectoid transformation. Needless to say, the positions of the curves differ from alloy system to alloy system and also with composition change within any specific alloy system. Examples of TTT curves for hypo- and hypereutectoid alloys of the iron-carbon system are presented in Fig. 5-9. There is in this chart an additional curve showing the appearance of the proeutectoid (ferrite) constituent.

It should be clear that the TTT diagram is in no sense a phase diagram; it has been introduced here merely to describe the nature of eutectoid transformation. The principles of equilibrium have no direct bearing upon transformation rates.

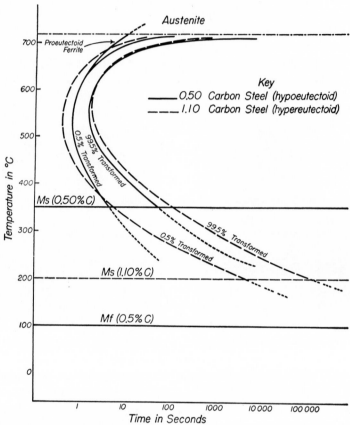

FIG. 5-9. Time-temperature-transformation diagram for two steels of different composition.

## Transformation at Various Cooling Velocities

Heat treatments involving isothermal subcritical reaction, such as those just described, are of recent origin and still somewhat unusual; in the case of steels this type of treatment is called *austempering*. Most of the common heat treatments such as quenching, normalizing, and annealing involve cooling at a more or less definite rate from above the eutectoid temperature. Thus, if the eutectoid alloy is cooled at a rate described by the path of the dashed line designated as *a* in Fig. 5-10, the $\gamma$ phase will be transformed to pearlite over a temperature range. The first pearlite formed at a high temperature will be coarser than that last formed at a

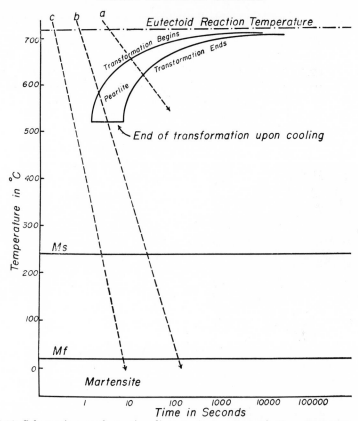

FIG. 5-10. Schematic transformation diagram for a eutectoid alloy. Dashed lines *a*, *b*, and *c* represent different cooling rates.

relatively lower temperature, so that a variation in the fineness of the pearlite will appear in the microstructure (Fig. 5-11). This type of transformation is typical of the annealing heat treatment, in which a very slow cooling rate is employed, or of normalizing, in which there is a somewhat faster rate of cooling obtained by permitting the metal to cool freely in the air.

A very fast rate of cooling, such as might result from quenching in cold water, is described by the dashed line *c* in Fig. 5-10. Here the transformation to $\alpha + \beta$ is suppressed altogether

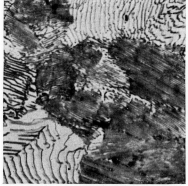

FIG. 5-11. Pearlite formed during cooling through the transformation range exhibits a wide variation in lamellar spacing. Magnification 1,500.

and the alloy is transformed wholly to martensite. In many alloys the $M_f$ line (completion of the transformation to martensite) lies below room temperature. When this happens, the conversion to martensite upon quenching to room temperature is incomplete. In either case, a tempering heat treatment, involving heating above the upper martensite line $M_s$, will cause a partial or complete transformation to $\alpha + \beta$, depending upon the time and temperature of tempering.

Curve $b$ represents an intermediate cooling rate. Here, transformation proceeds part way, is interrupted, and then the balance of the $\gamma$ is converted to martensite. The resulting microstructure is then composed partly of martensite (white) and partly of fine pearlite (gray) and a little bainite (dark) (Fig. 5-12).

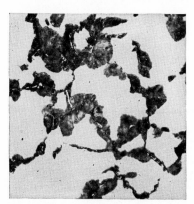

FIG. 5-12. Eutectoid steel partly transformed by cooling at an intermediate rate through the temperature range of transformation. Light areas are martensite; dark areas are partly pearlite and partly bainite. Magnification 500.

The diagram of Fig. 5-10 is entirely schematic, being intended to picture the general relationships that have just been described. For precision it would be required that the "cooling curves" (lines $a$, $b$, and $c$) show arrests (with "recalescence") and that adjustments be made in the TTT curves so that they might apply to cooling rather than to isothermal transformation. A quenching TTT diagram, obtained by noting the first traces of transformation upon cooling, lies a little to the right and below the isothermal TTT diagram for the same steel.

## Heat Treatment above the Eutectoid Temperature

The transformation of the $\alpha + \beta$ structure back to $\gamma$ at supereutectoid temperatures is a nucleation and growth process implemented by diffusion and, therefore, may be expected to proceed the faster the higher the temperature and the finer the $\alpha + \beta$ structure, i.e., the shorter the diffusion distance. In the case of the Fe-Fe$_3$C eutectoid, it is necessary to exceed the equilibrium transformation temperature by many degrees in order to achieve complete transformation to $\gamma$ within a reasonable time (austenitizing heat treatment).

Hypo- and hypereutectoid alloys must be heated correspondingly above the $\gamma$ boundaries of the $\alpha + \gamma$ and $\beta + \gamma$ fields, respectively, in order to convert wholly to the $\gamma$ phase. If the alloy is heated into and held within the range of two-phase equilibrium ($\alpha + \gamma$ or $\beta + \gamma$), first the eutectoid constituent will be converted to $\gamma$, and subsequently, as much of the proeutectoid $\alpha$ (or $\beta$, as the case may be) as is required to satisfy the requirements of equilibrium at the temperature of heat treatment will also be converted to $\gamma$. Upon subsequent cooling, only that part of the structure which had been converted to $\gamma$ will transform according to the mechanism described in the previous section. There will then be, in the structure, remnants of the original proeutectoid constituent plus products of decomposition of the $\gamma$ of a kind dictated by the cooling rate.

### Some Other Heat Treatments

Since eutectoid alloys are usually made by freezing from the melt, some nonuniformities, such as the coring that is characteristic of cast structures, are superimposed upon the normal eutectoid structure (Fig. 11-7). These may sometimes be reduced by a high-temperature heat treatment, wherein the temperature should be much higher than that required for the conversion to $\gamma$ and should be held for a long time, because diffusion over relatively long distances is required. The most favorable temperature for homogenization lies a few degrees below the solidus temperature.

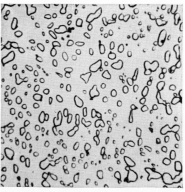

Spheroidization of the eutectoid constituent can be brought about by heating for an extended period at a temperature somewhat below that of the eutectoid (Fig. 5-13). No new principles are involved; spheroidization occurs by diffusion as in the case of eutectic alloys. This and the aforementioned homogenization treatment involve no phase changes, and the phase diagram is of service in guiding such treatments only in the information that it gives concerning temperature limits within which the treatments may be conducted.

FIG. 5-13. Eutectoid steel in which the $Fe_3C$ has been spheroidized by long heating at a temperature slightly below that of the eutectoid equilibrium; compare with Fig. 5-5$A$. Magnification 1,500.

### A Note Concerning the Stability of Iron-Carbon Alloys

The fact that the usual iron-carbon phase diagram (Fig. 5-2) represents metastable equilibrium between iron and iron carbide has already been mentioned. The *stable* system iron-graphite (Fig. 5-14) involves the same types of reactions on the iron side of the system, but with graphite substituted for $Fe_3C$ where the latter appears in the usual diagram. Reversion to the stable state, known as graphitization, proceeds only at relatively high temperatures and very slowly. Although this transformation is not reversible, the metastable state may be reestablished either by remelting and freezing or, if the alloy falls within the composition range of the $\gamma$ phase, by rehomogenization to $\gamma$ with subsequent cooling to produce $Fe_3C$ by eutectoid decomposition. From this it will be evident that references above to the formation of an $\alpha + Fe_3C$ eutectoid constituent by equilibrium transformation are not strictly correct; *true equilibrium transformation would yield a eutectoid constituent of $\alpha$ plus graphite.*

### Alloying by Isothermal Diffusion

There are a number of surface treatments that are applied to iron and its alloys which involve the diffusion of an alloying element into the

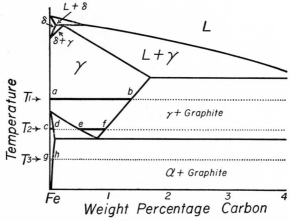

Fɪɢ. 5-14. Phase diagram of the stable system iron-carbon. Horizontal lines *ab*, *cdef* and *gh* are not part of the diagram; these have been inserted to indicate the composition ranges produced by carburizing treatment.

metal from an outside source. The principles of alloying by isothermal diffusion have been discussed in the preceding chapter, but some additional and important complexities appear in the eutectoid system iron-carbon owing to the metastability of $Fe_3C$.

The structures produced in this case by isothermal diffusion are those dictated by the stable system iron-graphite (Fig. 5-14), while structural evolution, accompanying the cooling of the $\gamma$ phase, that has been formed by diffusion is predicted by the metastable phase diagram (Fig. 5-2).

Three basically different diffusion structures may be produced by isothermally diffusing carbon into iron, depending upon the range within which the temperature of diffusion falls. Typical examples of each of these are designated in Fig. 5-14. At temperature $T_1$, above the $\alpha$ to $\gamma$ transformation in pure iron, only layers of $\gamma$ and carbon (graphite) can form, but the composition range within the $\gamma$ layer will extend from pure iron (point $a$) to saturation with carbon (point $b$) if the material is examined before equilibrium has been established. If a lower temperature $T_2$ is employed, three layers are formed, namely, $\alpha$, from $c$ to $d$; $\gamma$, from $e$ to $f$; and graphite. Here, as in previous examples, the $\alpha + \gamma$ and $\gamma +$ graphite regions of the phase diagram are represented in the diffusion sample simply as the interface between the layers of $\alpha$ and $\gamma$ and of $\gamma$ and graphite, respectively. Below the eutectoid temperature, at $T_3$, a two-layer structure is again formed: $\alpha$, from $g$ to $h$, and graphite. These, be it noted, are the structures existing at the temperature of isothermal diffusion; upon cooling to room temperature further changes occur.

Consider, for example, the sample that had been treated at $T_1$ (Fig. 5-14); upon cooling to room temperature, the alloyed surface layer will undergo metastable eutectoid transformation (Fig. 5-15). At the extreme surface, where the highest concentration of carbon exists, the material will be hypereutectoid (i.e., proeutectoid $Fe_3C$ and eutectoid

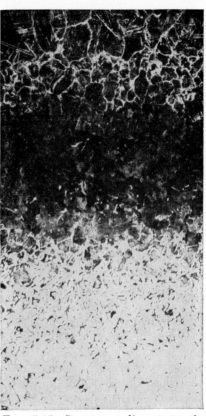

FIG. 5-15. Structure adjacent to the surface of a piece of carburized iron. External carbon-rich surface is at the top; unaffected iron at the bottom. Magnification 50.

$\alpha + Fe_3C$); somewhat below the surface the $\alpha + Fe_3C$ eutectoid will occur, and beneath that a hypoeutectoid zone. Thus, proeutectoid $Fe_3C$ rather than graphite will appear at the surface. At the same time proeutectoid $\alpha$ (ferrite) is being rejected adjacent to and merged with the pure iron central portion of the piece, which is also converted from $\gamma$ to $\alpha$. The zone of the eutectoid constituent, instead of being infinitely narrow, is usually of finite width (Fig. 5-15) because of undercooling with respect to the proeutectoid constituents over a composition range extending upon either side of the eutectoid point.

After cooling to room temperature, the structure of a piece of pure iron carburized at temperature $T_2$ fails to exhibit the sharp boundaries between layers that would have been apparent had it been possible to examine the material at the diffusion temperature. The layer of $\alpha$ remains virtually unchanged with cooling, but the $\gamma$ layer, adjacent to the surface, decomposes into hypereutectoid, eutectoid, and hypoeutectoid zones as in the previous example. The only obvious structural difference between this example and the one discussed previously lies in the narrower composition range represented in the decomposition products of the $\gamma$ layer.

The sample carburized below the eutectoid temperature at $T_3$ undergoes no significant structural change with cooling. At the diffusion temperature the metallic layer is composed solely of $\alpha$ (ferrite). Except for a very minor decrease in the solubility of carbon in $\alpha$ at lower temperatures, which can cause a slight precipitation of $Fe_3C$, the phase diagram does not predict any phase change for this case.

Exactly the reverse of the processes just described occurs if, instead of diffusing carbon *into* iron, carbon is diffused *out of* steel (decarburization). Additional factors are involved when other elements are diffused into or out of an iron-carbon alloy: these involve ternary alloy behavior, however, the principles of which will be discussed in a subsequent chapter.

### Limiting Cases of the Eutectoid

It will be obvious that the terminal solid-solution range of either the $\alpha$ or $\beta$ phase of the eutectoid reaction can be so narrow that it will appear to have zero width on the phase diagram. The eutectoid reaction isotherm will then appear to extend to the pure component. Occasionally the eutectoid point is located very close to one end of the reaction isotherm. The $\alpha + \gamma$ or the $\beta + \gamma$ field then becomes unrepresentably small and may be omitted in drawing the diagram. Where this happens, the user of the diagram must supply the missing region in his own mind in order to interpret the diagram correctly.

### Mechanical Properties of Eutectoid Alloys

The wide variety of structures which can be produced in alloys of eutectoid systems by controlled heat treatment yields a correspondingly great diversity in mechanical properties. No new principles beyond those that have already been stated are involved, but the relative importance of the several controlling factors differs from those obtaining with solid-solution and eutectic alloys. The principle of the dominance of the continuous phase still holds good, as does that of solid-solution hardening. However, the effect of particle size in two-phase mixtures, which was of minor consequence in previous examples, becomes a major factor in controlling the properties of eutectoid alloys, where the range of sizes is very large and extends to extremely fine particle size. The finer the particle size, the greater the hardness and strength. Hence, pearlite formed at low temperatures is harder than that formed at high temperatures, and the acicular $\alpha + \beta$ structures (bainite) are still harder. The transition phases, such as martensite, appear to be extremely hard, as a general rule. Accordingly, alloys quenched from the $\gamma$ range are the hardest of all. Tempering treatments, which cause the decomposition of the transition phase, reduce the hardness but increase the ductility, which normally increases as hardness decreases.

## PRACTICE PROBLEMS

**1.** Demonstrate that Fig. 5-1 has been drawn in compliance with the requirements of the phase rule.

**2.** Consider the alloy $X$ of Fig. 5-4, which upon normal cooling has transformed to a hypoeutectoid structure composed of proeutectoid $\alpha$ and eutectoid $\alpha + \beta$. If this alloy is next held for a long time at a temperature just below $T_3$, what changes will occur as the equilibrium state is approached? If this alloy were then quenched to room temperature and subsequently reheated to a temperature equal to one-half of $T_3$, what additional structural changes are to be anticipated?

**3.** A binary iron-carbon steel of eutectoid composition is subjected for a few minutes to decarburizing conditions at a temperature somewhat above that of the eutectoid but below the $\alpha$-$\gamma$ allotropic transformation temperature of pure iron and is then slowly cooled to room temperature. Deduce the microstructure of the affected zone in the steel.

# CHAPTER 6

# BINARY MONOTECTIC SYSTEMS

Another important three-phase reaction of the eutectic class is the *monotectic* (Fig. 6-1), in which one liquid phase decomposes with decreasing temperature into a solid phase and a new liquid phase.

$$L_\mathrm{I} \underset{\text{heating}}{\overset{\text{cooling}}{\rightleftharpoons}} \alpha + L_\mathrm{II}$$

Over a certain composition range, the two liquids are mutually immiscible, as are oil and water, and so constitute individual phases. Conformity of this type of system with the phase rule and other thermodynamic principles can be argued along lines parallel to those adopted in constructing the eutectic diagram. Accepted nomenclature follows the familiar pattern by which the meaning of such terms as *monotectic point, monotectic reaction isotherm, hypomonotectic,* and *hypermonotectic* will be understood. It should be noted, however, that the liquidus and solidus curves are differently located (Fig. 6-1) and that these have been designated as "upper" and "lower" to distinguish them. There is no special name for the boundary of the $L_\mathrm{I} + L_\mathrm{II}$ field; it will be called simply the

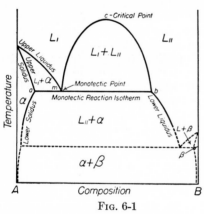

FIG. 6-1

*limit of liquid immiscibility.* The eutectic reaction, depicted by dashed lines in this example, is included merely to carry the diagram into the temperature range where all phases are solid. There are several other constructions that could as well have been used for this purpose.

## Monotectic Decomposition

Above the *monotectic temperature,* the *monotectic alloy,* $Y$ in Fig. 6-2, is composed solely of a single liquid phase $L_\mathrm{I}$. At (or slightly below) the monotectic temperature, this liquid decomposes in a manner entirely

72

analogous to the decomposition of the eutectic liquid, the only difference being that one of the products is a liquid instead of a solid phase. Thus, the resulting monotectic constituent resembles the eutectic constituent in all respects except that one of the phases is fluid. As it happens, all known

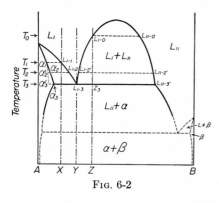

FIG. 6-2

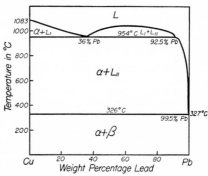

FIG. 6-3. Phase diagram of the system copper-lead.

binary monotectic points in metal systems are located nearer to the composition of the solid phase than to that of the second liquid so that the solid phase predominates in the monotectic constituent.

$\%\alpha$ (monotectic)

$$= \frac{L_{I-3}L_{II-3}'}{\alpha_3 L_{II-3}'} \times 100 \approx 75\%$$

$\%L_{II}$ (monotectic)

$$= \frac{\alpha_3 L_{I-3}}{\alpha_3 L_{II-3}'} \times 100 \approx 25\%$$

Accordingly, the liquid $L_{II}$ appears in pockets apparently surrounded by the solid $\alpha$ phase (Fig. 6-4). This is a picture of a copper-lead monotectic alloy (Fig. 6-3) and was, of course, taken after the alloy had cooled to room temperature. The liquid $L_{II}$ had meanwhile frozen as a eutectic constituent composed of 99.5% lead and 0.5% copper and appears as gray areas in the picture.

Because of the continuity of the copper-rich $\alpha$ phase in the monotectic

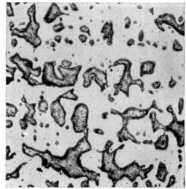

FIG. 6-4. Microstructure of cast monotectic alloy Cu + 36% Pb. Light areas are the Cu-rich matrix of the monotectic constituent; dark areas are the Pb-rich portion, which existed as $L_{II}$ at the monotectic temperature. Magnification 100.

structure, the physical properties of this alloy more nearly resemble those of copper than those of lead. Lead is often added to alloys in which it introduces monotectic reaction because it makes the machining of the metal easier, by reducing the ductility just enough to cause chips to break away, without seriously decreasing hardness and strength. Leaded alloys are used also for bearings, where the continuous phase of the high-melting metal gives strength to the member, while the lead, occurring in pockets at the running surface, serves to reduce the friction between bearing and axle.

## Hypomonotectic Alloys

Alloys on the "$\alpha$ side" of the monotectic point, the *hypomonotectic alloys*, composition $X$ in Fig. 6-2, for example, pass through a state of bivariant equilibrium before the univariant monotectic reaction occurs. Freezing begins with a primary separation of the $\alpha$ phase, which may be cored because of failure to maintain equilibrium in the $\alpha$ phase: $\alpha_1$, $\alpha_2'$, $\alpha_3'$. At the monotectic temperature $T_3$ there will be some excess of $L_{I-3}$ over the equilibrium quantity, and this will react to yield the secondary monotectic constituent. Thus, the hypomonotectic copper-lead alloy of Fig. 6-5 is composed of cored copper-rich primary dendrites interlaid with a divorced monotectic composed of $\alpha_3$ continuous with the primary $\alpha$ dendrites and $L_{II-3}'$ in the interdendritic spaces. According to the copper-lead diagram (Fig. 6-3) the $L_{II}$ in the pockets between the $\alpha$ dendrites should, with further cooling below the monotectic temperature, deposit a tertiary $\alpha$ phase, which will be physically continuous with and indistinguishable in the picture from the primary and secondary $\alpha$. Finally, at the eutectic temperature the $L_{II}$ will decompose into a fourth-order constituent, the eutectic of copper-rich $\alpha$ and lead-rich $\beta$. This eutectic, being of very fine particle size, is not resolved in the photomicrograph and appears as dark spots where the $L_{II}$ had been located.

## Hypermonotectic Alloys

The occurrence of two-liquid immiscibility in the hypermonotectic alloys at temperatures above the monotectic introduces structural considerations which have no parallel in the examples discussed up to this point. Given time enough, the two liquids will separate into two layers placed according to density, with the lighter layer on top, as oil floats on water. It is quite possible, however, to have the two liquids existing as an emulsion, wherein tiny droplets of one liquid remain suspended in the other liquid. Which of these conditions will be the more nearly approached in a given heating and cooling cycle must depend upon the physical characteristics of the alloy system concerned, upon the conditions of formation of the second liquid, and upon the opportunity afforded for the segregation of the two liquids. Knowledge of these matters with respect to metals is fragmentary and unsatisfactory at the present time.

Suppose that a hypermonotectic alloy, such as $Z$ in Fig. 6-2, is heated to a high temperature above $T_0$; the alloy will be composed of one liquid phase. When, upon cooling, the limit of liquid immiscibility is crossed at $L_{I-0}$, the second liquid $L_{II-0}$ should make its appearance, probably at the surface of the confining vessel and possibly, also, at various points

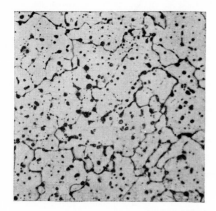

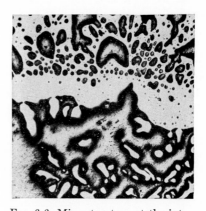

FIG. 6-5. Hypomonotectic Cu + 5% Pb alloy, as cast. Light areas are primary and monotectic Cu; dark areas are Pb which existed as $L_{II}$ at the monotectic temperature. Magnification 100.

FIG. 6-6. Microstructure at the interface between the two layers in a Cu + 50% Pb hypermonotectic alloy. Upper layer is the product of the monotectic decomposition of $L_I$ (36% Pb); lower layer contains light particles of Cu, precipitated during the cooling of $L_{II}$ from the monotectic temperature down to that of the eutectic, embedded in the dark Pb-rich Pb-Cu eutectic constituent. Magnification 100.

throughout the liquid bath. As the temperature falls, the quantity of $L_{II}$ increases, so that just above the monotectic temperature its amount will be

$$\%L_{II} = \frac{L_{I-3}z_3}{L_{I-3}L_{II-3}'} \times 100 \approx 20\%$$

Conditions being favorable, this liquid will exist as a separate layer in the crucible or mold. That portion of the mixture which is composed of $L_{I-3}$ will now react to form the monotectic constituent. In the case of the copper-lead (50-50) alloy in Fig. 6-6, this is the upper of the two layers; the lower layer of $L_{II-3}'$ remains unchanged during this reaction. Now the alloy is composed of an upper layer in which pockets of $L_{II}$ are surrounded by the $\alpha$ phase and a lower layer of $L_{II}$, which is still fully molten. With continued cooling the $L_{II}$ in both layers rejects a small quantity of the $\alpha$ phase in descending to the eutectic temperature and then freezes as a

eutectic constituent of $\alpha + \beta$. The two layers and their respective structures are apparent in Fig. 6-6.

### The Field of Two-liquid Immiscibility

As with all two-phase equilibria, the field of $L_I + L_{II}$ should be regarded as being made up of an infinite number of tie-lines connecting, at
each temperature, the composition of $L_I$ that can exist at equilibrium with
the conjugate composition of $L_{II}$. This region has been closed at the top
by the meeting of the two lateral boundaries, as it is drawn in Figs. 6-1
and 6-2 and as shown in detail in Fig. 6-7. With increasing temperature
the tie-lines become progressively shorter until the ultimate tie-line, at
the top of the area, has zero length. This is a critical point $c$ where the two

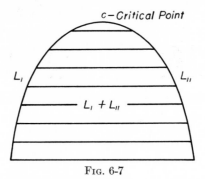

FIG. 6-7

liquid phases, having identical composition, become indistinguishable;
i.e., the meniscus between them vanishes. At higher temperature, only
one liquid solution exists over the entire composition range of the alloy
system.

Two-liquid equilibria terminating in critical points have been shown to
exist, but it is worth mentioning that this construction has been included
in a number of the published phase diagrams where the experimental observations needed to justify the inclusion of a critical point are lacking.
The two-liquid region can terminate, also, at high temperature on another
three-phase isotherm, such, for example, as one involving the two liquids
and the gas phase (Chap. 11).

### Limiting Cases of the Monotectic

Very limited ranges of terminal solid solubility are found in most alloy
systems of the monotectic type. This seems reasonable, because any
tendency to segregate into separate phases in the liquid state might be
expected to be even more pronounced in the solid state. As a result of the

narrow ranges of solid solubility, coring and accompanying effects tend to be less pronounced in monotectic than in eutectic systems.

There are many combinations of metals which are thought to be virtually insoluble in both the liquid and solid states. This condition corresponds to a limiting case of the monotectic, combined with the eutectic (or peritectic) reaction (see Fig. 6-8). The upper of the two horizontal lines represents a monotectic reaction in which the monotectic point $L_I$ is

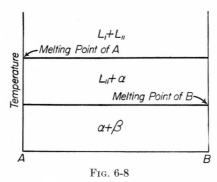

FIG. 6-8

almost coincident with the melting point and composition of pure $A$, and $L_{II}$ is nearly coincident with the composition of pure $B$. The eutectic point, on the lower of the two horizontal lines, is, similarly, nearly coincident with the melting point and composition of pure $B$. Thus, the two metals when melted in the same container appear to melt at their individual melting points and to remain as separate layers throughout melting and freezing.

### PRACTICE PROBLEMS

**1.** Draw a phase diagram of a monotectic type system (similar to that shown in Fig. 6-1) where $A$ is the higher melting metal but the monotectic point $m$ occurs at 75% $B$ (instead of at 25% $B$). How will this change affect the relative proportions of the products of monotectic transformation? How might this affect the freezing behavior of the monotectic alloy?

**2.** In Fig. 6-2, consider an alloy of composition midway between pure $A$ and composition $X$. Deduce the cast microstructure of this alloy. How will the cast microstructure be altered by heating for a long time at temperature $T_3$? If then this alloy is slowly cooled, what sequence of further changes will occur?

**3.** With reference to Fig. 6-7, consider three alloys having compositions, respectively, (a) slightly to the left of point $c$, (b) coincident with point $c$, and (c) slightly to the right of point $c$. Trace the course of transformation of each of these alloys upon cooling, and compare them.

# CHAPTER 7

# CONGRUENT TRANSFORMATION OF ALLOYS

When one phase changes directly into another phase without any alteration in composition during the transformation, the phase change is said to be congruent. Conversely, an *incongruent phase change* is one which requires either a transient or a persistent composition change, as in the freezing of the solid solution or of the eutectic-type alloys, respectively. All pure components transform congruently. An example of the congruent melting of an alloy was presented at the end of Chap. 3, where it was pointed out that the minimum (or maximum) melting point in an isomorphous system lies at a point of congruent melting.

The *intermediate phases*, which are those that are not isomorphous with either of the components of the alloy system (i.e., they occur between the terminal phases), are often classified in two groups according to whether they are congruently melting or incongruently melting. The incongruently melting intermediate phases will be considered in succeeding chapters, where the peritectic, peritectoid (incongruently transforming), and syntectic reactions are discussed. For the moment, attention will be centered upon the congruently melting phases. These are sometimes referred to as *intermetallic compounds*. Although the use of this term is well justified in some instances upon the ground that a specific intermediate phase occurs at a composition corresponding to a simple ratio of the two kinds of atoms concerned, there are so many apparent exceptions that it seems better, on the whole, to regard all such phases simply as "intermediate phases" and to designate them by the use of Greek letters instead of molecular symbols. This practice has been adopted with increasing regularity in recent years by authors of phase diagrams.

An example of a congruently melting intermediate phase occurs in the magnesium-silicon system (Fig. 7-1). Here the $\beta$ phase (which, incidentally, corresponds to the composition $Mg_2Si$) divides the phase diagram into two independent portions. A eutectic system is formed by the alloys of magnesium and $\beta$, and another eutectic system is formed by the alloys of $\beta$ and silicon. Thus, it is seen that the congruently melting intermediate phase behaves as a *component*. Indeed, any intermediate phase which behaves congruently in all transformations to which it is subject may be

78

regarded as a component. This is the reason for defining the component (Chap. 1) in such manner as not to limit it to elemental substances. In the binary systems there is no particular advantage to be derived from splitting the phase diagrams into sections lying between terminal and intermediate "components." In ternary alloy systems, however, a degree of simplicity is gained by this procedure, as will be demonstrated in Chap. 17.

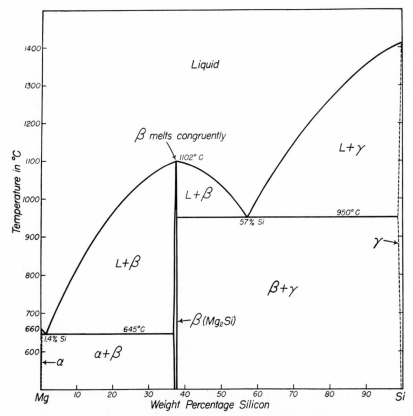

FIG. 7-1. Phase diagram of the system magnesium-silicon.

It will be evident that the example presented is only one of many possible combinations. The occurrence of two eutectics in this system is coincidental. Any type of equilibrium can occur among the several components. It should be noted also that the composition range of the intermediate phase may, in some instances, be very narrow, so that it can be represented only by a single line. As a matter of fact, the width of the $\beta$ field has been exaggerated in Fig. 7-1 to permit it to be represented as an area. Broad solid-solution ranges associated with intermediate phases are also found.

### Congruent Transformation in the Solid State

Congruent transformation may occur with solid-solid as well as solid-liquid phase changes. An interesting example is found in the iron-chromium system (Fig. 7-2). Here, three kinds of congruent transformation are represented, namely, (1) a minimum melting point at $a$, (2) congruent transformation of the $\gamma$ terminal solid solution to the $\alpha$ solid solution at a

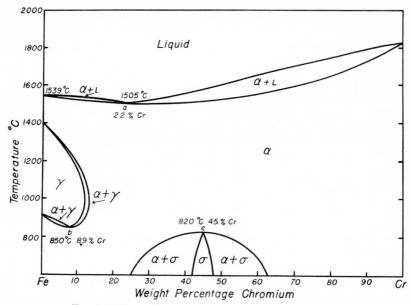

FIG. 7-2. Phase diagram of the system iron-chromium.

minimum transformation point $b$, and (3) congruent transformation of the $\sigma$ intermediate solid-solution phase to $\alpha$ at a maximum transformation point $c$. In this case the only fully congruent alloy is that corresponding to the composition of the minimum melting point; the "$\sigma$ alloy" and the "$\gamma$ alloy" melt incongruently. Hence, the division of this diagram into two individually complete parts is possible only at the minimum melting composition. It is worth noting also that the so-called "$\gamma$ loop" does not pass through a minimum or maximum in all alloy systems in which it is found. The $\alpha + \gamma$ region may, alternately, be formed as a crescent with its maximum and minimum points occurring at the pure metal composition.

## Ordered Crystal Structure

Many of the intermediate phases have ordered crystal structures. This means that the two kinds of atoms, instead of being located at random on the crystal lattice as in ordinary terminal solid solutions of the substitutional type, have specific positions on the crystal lattice. In a simple case, for example, the two kinds of atoms might alternate along any principal direction in the crystal lattice, so that in space each $A$ atom would have only $B$ atoms as nearest neighbors and each $B$ atom would have only $A$ atoms as nearest neighbors.

This detail of crystal structure would not require special comment were it not for the fact that there are a few cases where an ordered crystalline variety transforms (congruently) into a disordered crystalline variety that is crystallographically identical except for the absence of ordering.[1] For this reason, the question has been raised whether or not such transformation should be regarded as a true phase change.

The subject of heterogeneous equilibrium, however, does not deal with the structure of the phases per se; it is concerned only with the conditions of coexistence of distinguishable states of matter. To prove a difference in phase, it is sufficient to show that at equilibrium, the two states can coexist in physical contact with a sharp boundary between them. Criteria for distinguishing two states so situated may be any externally measurable property. Recent studies[2] have demonstrated, for the contested cases of $Cu_3Au$ and some others, that the ordered and disordered states can coexist at equilibrium. Thus, it is correct to represent the order-disorder transformation upon the phase diagram as a normal phase change.

## Physical Properties of Intermediate Phases

Ductility and softness are usually associated with simple crystal types, such as those of most of the common metals and their associated terminal solid solutions. Crystals of complex structure usually exhibit great hardness but are relatively brittle and often fragile. Since the majority of intermediate phases have complex crystal structures, it is to be expected that they will tend to be hard and brittle. There are, of course, noteworthy exceptions. Eutectic, eutectoid, and other two-phase constituents involving an intermediate phase naturally partake of the properties of that phase in accordance with its proportion and distribution. Hence, multiphased alloys that include intermediate phases are, as a class,

---

[1] For a detailed review of this subject, see F. C. Nix and W. Shockley, Order-Disorder Transformations in Alloys, *Revs. Mod. Phys.* **10**:1 (1938).

[2] F. N. Rhines and J. B. Newkirk, The Order-Disorder Transformation Viewed as a Classical Phase Change, *Trans. Am. Soc. Metals*, **45**:1029–1055 (1952).

harder and more brittle than those composed of terminal solid solutions alone.

## PRACTICE PROBLEMS

**1.** Draw a phase diagram for the system $AB$, which has an intermediate phase $\gamma$ at 50% $B$; the partial system $A\gamma$ is of the eutectic type; the partial system $\gamma B$ is of the monotectic type at high temperature with a eutectic reaction at lower temperature.

**2.** Deduce the structural changes that will occur in cooling an (8.9%) iron-chromium alloy from 1600°C to room temperature. What difference would be found if the composition were changed to 8.0% chromium? (Note: use Fig. 7-2 in answering this problem.)

**3.** Iron-chromium alloys (Fig. 7-2) composed of 44 and 45% chromium, respectively, will both transform from $\alpha$ to $\sigma$ upon cooling from 1000°C to room temperature. What differences should be observed in the paths of transformation of these two alloys?

CHAPTER 8

# BINARY PERITECTIC SYSTEMS

Corresponding to the eutectic group of univariant reactions is a group of *peritectic*-type reactions, in which one phase decomposes with *rising* temperature into two new phases. It will be recalled that the eutectic-type reactions are all characterized by the decomposition with *falling* temperature of one phase into two new phases. The peritectic reaction itself consists, upon heating, in the decomposition of one solid phase into a liquid and a new solid phase:

$$L + \alpha \underset{\text{heating}}{\overset{\text{cooling}}{\rightleftharpoons}} \beta$$

In other words, the solid phase melts *incongruently*, i.e., with decomposition. This type of reaction is represented in Fig. 8-1, where the $\beta$ phase of peritectic composition (point $p$) melts by decomposition into $L + \alpha$ of

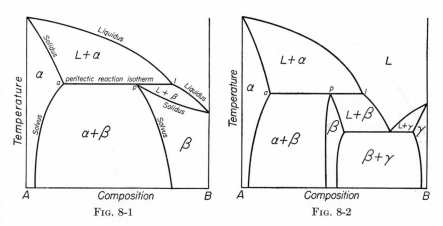

Fig. 8-1                    Fig. 8-2

compositions $l$ and $a$, at the peritectic reaction temperature. It is not necessary that the solid phase at the peritectic point be a terminal solid solution; indeed, it is more often an intermediate phase (Fig. 8-2) that melts incongruently.

All the arguments required to derive the method of representation of peritectic equilibrium from the phase rule have been cited in connection

83

with the eutectic systems and need not be repeated here. The peritectic line must, of course, be horizontal (isothermal) because it represents univariant equilibrium among three phases. Where they intersect the peritectic reaction isotherm, all boundaries of two-phase regions must arrive at such angles that they project into other two-phase regions and not into one-phase regions of the diagram. Any one-phase region may touch the peritectic line at only one single composition point.

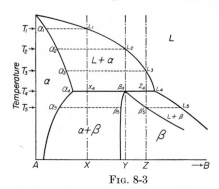

FIG. 8-3

### Equilibrium Freezing of Peritectic Alloys

Consider the course of freezing of the *peritectic alloy*, composition $Y$ in Fig. 8-3. Upon cooling from the fully molten condition, the liquidus is reached at temperature $T_2$, where crystals of the $\alpha$ phase ($\alpha_2$) begin to form. The liquid composition is displaced to the right as the temperature falls, and more solid is deposited. The composition of the $\alpha$ phase changing along the solidus reaches the end of the *peritectic line* ($\alpha_4$) when the liquid composition has reached the opposite end ($L_4$).

$$\%\alpha \text{ (alloy } Y \text{ above peritectic)} = \frac{\beta_4 L_4}{\alpha_4 L_4} \times 100 \approx 35\%$$

$$\%L \text{ (alloy } Y \text{ above peritectic)} = \frac{\alpha_4 \beta_4}{\alpha_4 L_4} \times 100 \approx 65\%$$

Peritectic reaction now occurs; the liquid and $\alpha$ react together to form $\beta$. Under "equilibrium" conditions freezing must be completed isothermally by this process, and all the previously formed $\alpha$ as well as the liquid must be consumed.

If the alloy composition had lain to the left of the peritectic point, as at $X$ in Fig. 8-3, the $\alpha$ phase would not have been totally consumed in forming $\beta$. In this alloy, freezing begins at $T_1$, where $\alpha_1$ first appears. Just above the peritectic temperature the quantities of $\alpha$ and liquid will be

$$\%\alpha \text{ (alloy } X \text{ above peritectic)} = \frac{x_4 L_4}{\alpha_4 L_4} \times 100 \approx 80\%$$

$$\%L \text{ (alloy } X \text{ above peritectic)} = \frac{\alpha_4 x_4}{\alpha_4 L_4} \times 100 \approx 20\%$$

Just below the peritectic temperature the liquid will have disappeared and there will be only the solid phases $\alpha$ and $\beta$.

$$\%\alpha \text{ (alloy } X \text{ below peritectic)} = \frac{x_4\beta_4}{\alpha_4\beta_4} \times 100 \approx 70\%$$

$$\%\beta \text{ (alloy } X \text{ below peritectic)} = \frac{\alpha_4 x_4}{\alpha_4\beta_4} \times 100 \approx 30\%$$

Notice that the quantity of the $\alpha$ phase has decreased during peritectic reaction from about 80 to about 70%.

Had the composition of the alloy lain to the right of the peritectic point, alloy $Z$ in Fig. 8-3, for example, then an excess of liquid would have survived the peritectic reaction. Freezing in this case begins at $T_3$, where $\alpha_3$ is first rejected. There is a small quantity of the $\alpha$ phase present by the time the peritectic temperature is reached:

$$\%\alpha \text{ (alloy } Z \text{ above peritectic)} = \frac{z_4 L_4}{\alpha_4 L_4} \times 100 \approx 10\%$$

$$\%L \text{ (alloy } Z \text{ above peritectic)} = \frac{\alpha_4 z_4}{\alpha_4 L_4} \times 100 \approx 90\%$$

But this is entirely consumed in forming the $\beta$ phase:

$$\%\beta \text{ (alloy } Z \text{ below peritectic)} = \frac{z_4 L_4}{\beta_4 L_4} \times 100 \approx 30\%$$

$$\%L \text{ (alloy } Z \text{ below peritectic)} = \frac{\beta_4 z_4}{\beta_4 L_4} \times 100 \approx 70\%$$

Observe that the quantity of liquid has been reduced to about 70% and that the quantity of $\beta$ formed is much greater than that of the $\alpha$ originally present.

## Natural Freezing of Peritectic Alloys

The departure from equilibrium in the natural freezing of peritectic alloys is usually very large. This is because the reaction product, the $\beta$ phase, necessarily forms at the interface between $\alpha$ and liquid, establishing a hindrance to the diffusion which is necessary for the continuation of the reaction (see Fig. 8-4). In order to change $\alpha$ into $\beta$, some $A$ atoms must diffuse out of the $\alpha$ phase and are replaced by $B$ atoms, which must come, ultimately, from the liquid phase, both having been trans-

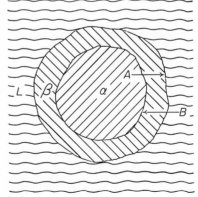

FIG. 8-4. Schematic representation of primary $\alpha$ undergoing peritectic reaction with liquid $L$ to form an envelope of $\beta$. The reaction progresses by the diffusion of $A$ atoms outward and $B$ atoms inward through the shell of $\beta$.

ported through the zone of $\beta$. As the $\beta$ layer grows thicker, the distance over which the two kinds of atoms must be transported increases. Thus, the first layer of $\beta$ forms quickly, but its growth proceeds at a decelerating pace, so that the reaction is frequently incomplete when the temperature has fallen well below that of the peritectic and, indeed, has become so low that further diffusion is negligible.

When the entire process of natural freezing is considered, it is found that further deviation from equilibrium is caused by the familiar coring effect. Alloy $X$ of Fig. 8-5 begins freezing by the deposition of a cored

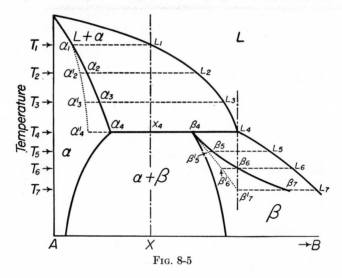

Fig. 8-5

primary $\alpha$ constituent. Hence, at the peritectic temperature, the quantity of the $\alpha$ is less and that of the liquid larger than should be the case under conditions of equilibrium:

$$\%\alpha'_4 \text{ (alloy } X \text{ above peritectic)} = \frac{x_4 L_4}{\alpha'_4 L_4} \times 100 \approx 60\%$$

$$\%L_4 \text{ (alloy } X \text{ above peritectic)} = \frac{\alpha'_4 x_4}{\alpha'_4 L_4} \times 100 \approx 40\%$$

Also, the primary $\alpha$ constituent is richer in the $A$ component than it would otherwise be. Partial reaction of liquid with the $\alpha$ phase now occurs. A layer of $\beta$, of peritectic composition, is formed at the expense of some of the $\alpha$ and a larger share of the liquid. When, as usually happens, the temperature falls below $T_4$ before peritectic reaction is complete, the remaining liquid, of composition $L_4$, will freeze directly to cored $\beta$; that is, $L_4$ deposits $\beta_4$ first, then successive $\beta$ compositions along the lower solidus, yielding average $\beta$ compositions of $\beta'_5$, $\beta'_6$, and $\beta'_7$. At $T_7$ the liquid is finally consumed and the alloy is entirely solid. The microstructure now consists of a primary cored $\alpha$ constituent surrounded by zones of

Fig. 8-6. Schematic representation of "envelopment" in a cast peritectic-type alloy. A residue of cored primary $\alpha$ is represented by the solid circles concentric about smaller dashed circles; surrounding the cored $\alpha$ is a layer of $\beta$ of peritectic composition, the remaining space being filled with cored $\beta$, represented by dashed curved lines.

secondary peritectic $\beta$ of constant composition, with tertiary cored $\beta$ filling the remaining space (Fig. 8-6).

## Envelopment

Frequently, the layer of peritectic $\beta$ is so thin that it is not perceptible in the microstructure. When it is sufficiently thick to be seen, however, it appears in the form of "envelopes" surrounding each of the particles of

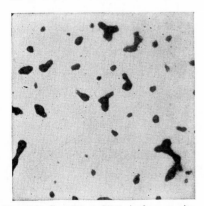

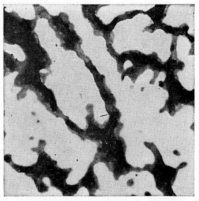

Fig. 8-7. Cast Pt + 25% Ag hypoperitectic alloy. White and light gray matrix is primary cored $\alpha$ (Pt-rich); dark spots are $\beta$. Magnification 500.

Fig. 8-8. Cast Pt + 60% Ag hyperperitectic alloy. White and light gray areas are residual cored $\alpha$; dark two-toned areas are $\beta$, the outer portions being of peritectic composition and the darkest central areas being the cored $\beta$ that formed at temperatures below that of the peritectic reaction. Magnification 1,000.

the primary constituent. This structural feature bears the name *envelop-* *ment*. Examples of envelopment appear in Figs. 8-7 and 8-8, which depict the microstructures of two alloys of the peritectic system silver-platinum (Fig. 8-9). Coring of the light-colored $\alpha$ phase is shown by a change from white to light gray; $\beta$ envelopment is seen as a medium-gray layer between the light and dark areas; the darkest areas are the cored $\beta$.

### Freezing of Some Other Alloys

An alloy, such as $Z$ in Fig. 8-10, which, under equilibrium conditions, would freeze to a structure composed of $\beta$ alone, may develop a two-phase structure through the process of natural freezing. Primary cored $\alpha$ crystals separate in the temperature range $T_1$ to $T_4$. If peritectic reaction is incomplete at $T_4$, some of this primary $\alpha$ constituent may remain enveloped

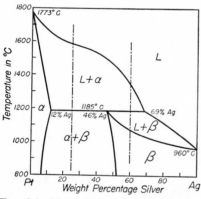

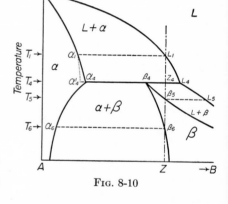

FIG. 8-9. Phase diagram of the system platinum-silver.

FIG. 8-10

by secondary peritectic $\beta$ as is seen to have occurred in the 60% silver alloy of Fig. 8-8. The remaining liquid freezes to cored $\beta$. Further changes in the solid state will be discussed presently.

The reverse of this effect is found in alloys beyond the end of the peritectic isotherm, on the side toward the solid ($\alpha$) phase, alloy $W$ in Fig. 8-11. Here, equilibrium freezing would produce a structure composed wholly of the $\alpha$ phase, but natural freezing develops some of the $\beta$ phase in the cast structure. Coring of the primary $\alpha$ constituent being sufficient, some liquid will remain at $T_4$:

$$\%\alpha \text{ (alloy } W \text{ above peritectic)} = \frac{W_4 L_4}{\alpha_4' L_4} \times 100 \approx 90\%$$

$$\%L \text{ (alloy } W \text{ above peritectic)} = \frac{\alpha_4' W_4}{\alpha_4' L_4} \times 100 \approx 10\%$$

This liquid will react with $\alpha$ to form peritectic $\beta$, and if it is not thus consumed, it will freeze directly to cored $\beta$.

Evidently, nonequilibrium freezing can result in an excess of either the $\alpha$ or the $\beta$ phase in the microstructure. Alloys rich in $A$ tend to have an excess of the $\beta$ phase, while those rich in $B$ tend to have an excess of the $\alpha$ phase. Differing relative diffusion rates in the several phases and differing constitutional relationships, with regard to the compositions of the phases on the peritectic line, from alloy system to alloy system can have a pronounced influence upon the character of the cast microstructure.

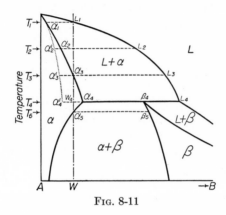

FIG. 8-11

It should be noted that although the eutectic and peritectic structures both involve two phases, the two types of microstructure may be distinguished readily by the facts that (1) the peritectic has no *two-phase constituent* and (2) both phases of the peritectic alloy may be cored.

### Heat Treatment of Peritectic Alloys

Where there exists coring inhomogeneity, or nonequilibrium proportions of the phases present, a homogenizing heat treatment, at a temperature such as to ensure an appreciable rate of diffusion without causing a phase change, may be expected to cause the structure of the alloy to change in the direction of equilibrium. Thus, alloy $W$ of Fig. 8-11 can be converted to homogeneous $\alpha$ by heating for some time at a temperature between $T_3$ and $T_5$; rapid cooling (quenching) to room temperature should, in general, cause the homogeneous $\alpha$ structure to be retained as a supersaturated solid solution. At any temperature where the diffusion rate becomes appreciable, up to $T_5$, however, the $\beta$ phase may reappear as a Widmanstätten precipitate. Age hardening is possible in this type of alloy just as in those alloy systems of the eutectic group where the solvus inclines toward lower solubility at low temperature.

In like manner, alloy $Z$ of Fig. 8-10 could be homogenized to uniform $\beta$ if it were heated for a long time at some temperature between $T_5$ and $T_6$. Slow cooling or quenching followed by reheating to a temperature below $T_6$ might be expected to cause the appearance of a Widmanstätten precipitate of the $\alpha$ phase. This alloy, too, might be subject to age hardening if the essential crystallographic relationships are present.

Intermediate alloys, such as $X$ in Fig. 8-5, cannot be homogenized to a single phase, but the relative proportions of the two phases may be adjusted toward the equilibrium ratio by heat treatment. Coring within the individual constituents may be eliminated, and spheroidization of the minor constituent may occur.

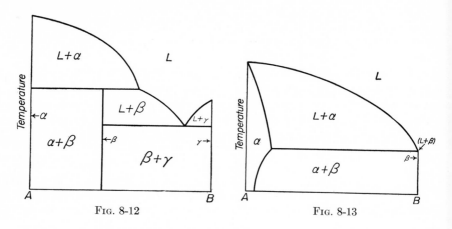

FIG. 8-12                    FIG. 8-13

Liquation occurs as in eutectic alloys, but with the difference that melting may be encountered at temperatures below that of the peritectic. If any of the cored $\beta$ constituent is present, and it may be present even in alloys to the left of the peritectic line, then liquation can be encountered at the fusion temperature of the lowest melting zone in the $\beta$ constituent. For this reason, a slow approach to the desired homogenizing temperature is especially important when dealing with peritectic alloys.

### Limiting Cases of the Peritectic

Where the solid-solution range of the $\beta$ intermediate phase becomes very narrow (Fig. 8-12), it may appear upon the phase diagram as a single line. Likewise the terminal solid-solution ranges become very narrow in some systems, so that no areas corresponding to the solid phases can be shown. These conditions influence the interpretation of the phase diagram only to the extent that the coring effect becomes negligible; incomplete peritectic reaction persists in such systems.

If the peritectic point falls very close to one end of the peritectic reaction isotherm, it may become difficult to distinguish between a eutectic and a peritectic system; compare Fig. 8-13 with Fig. 4-22. The only means of distinguishing between the two cases, in some instances, has been to make a very careful comparison of the reaction temperature with the melting point of the $B$ component (see Chap. 21). If the reaction temperature is higher, the system is peritectic; if lower, it is eutectic. In either case the $L + \beta$ field becomes so small that it cannot be shown.

## Mechanical Properties of Peritectic Alloys

Peritectic alloys, being usually two-phase aggregates, have mechanical properties that may be associated with microstructure in accordance with the principles stated for eutectic alloys. Two special characteristics of peritectic alloys should, however, be noted in this connection, namely: (1) the individual phases are ordinarily far removed from their respective states of homogeneous equilibrium and (2) the particle size of the cast peritectic alloy is rarely fine. The first of these characteristics means that a regular (linear) change in properties across the range of two-phase alloys is not to be expected unless special steps, such as working and heat treatment, have been taken to establish a state of near equilibrium. The coarseness of the peritectic structure is reflected in the physical properties only in a negative sense; that is, hardening by the deliberate refinement of the reaction products is very limited.

### PRACTICE PROBLEMS

**1.** The system $AB$ forms a congruently melting compound $AxBy$; alloys between $A$ and $AxBy$ are of the eutectic type; those between $AxBy$ and $B$ are of the peritectic type. Draw a phase diagram for this system.

**2.** Considering an alloy such as $X$, Fig. 8-5, how would the cast microstructure be altered by moving the peritectic point $\beta_4$ a little to the left? to the right?

**3.** Consider that alloy $W$, Fig. 8-11, was made homogeneous by long heat treatment at $T_4$. Then, just before cooling to room temperature, the heat-treating temperature was increased briefly to $T_2$. Deduce the resulting microstructure of alloy $W$.

**4.** Deduce the cast microstructure of a hyperperitectic alloy (that is, having its composition midway between $p$ and $l$) in the system illustrated in Fig. 8-2.

# CHAPTER 9

# BINARY PERITECTOID SYSTEMS

The *peritectoid reaction* (also known as the metatectic reaction) is related to the peritectic in much the same way as the eutectoid to the eutectic. Only solid phases are involved in peritectoid reaction:

$$\alpha + \gamma \underset{\text{heating}}{\overset{\text{cooling}}{\rightleftharpoons}} \beta$$

The solid phase $\beta$, upon heating, decomposes into two new solid phases (Fig. 9-1). This is one kind of incongruent transformation in the solid state.

Although peritectoid reaction is fairly common among the alloy systems at large, so few of the specific alloys of commerce occur within composition ranges involving peritectoid reaction that experience with this transformation has been very scanty. Upon the basis of its similarity to the peritectic it might be expected that the peritectoid transformation would be very slow. Reaction between the high-temperature phases ($\alpha$ and $\gamma$) during cooling is expected to proceed at the phase interface, where a layer of the low-temperature phase ($\beta$) should be formed. The continuation of the transformation would require diffusion through a growing septum of $\beta$ and ordinarily at a relatively low temperature where diffusion rates are small. Thus, it should be common to find the transformation incomplete when the alloy has reached room temperature.

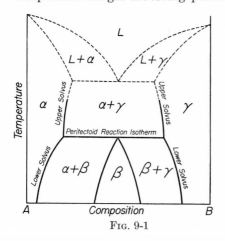

Fig. 9-1

Such experimental evidence as exists indicates the presence of several complicating factors. First, the transformation may proceed through a series of unstable transition states, which modify its mechanism and rate. Second, crystallographic factors, which influence the directions of diffusion and of phase growth, can affect the morphology of the reaction products. And finally, it frequently happens that the composition of the low-temperature phase ($\beta$) differs so little from the composition of one of the high-temperature phases that the impeding influence of diffusion upon the progress of transformation is minimized.

An example of the peritectoid is to be found in the silver-aluminum system (Fig. 9-2). Here the composition range of the $\beta$ phase partially overlaps that of the high-temperature phase $\gamma$. Quenched from a temperature slightly above that of the reaction, the microstructure of a 7% aluminum alloy is composed of a matrix of $\gamma$ with a few particles of $\alpha$ Fig. 9-3$A$). When this alloy has been cooled to a temperature slightly below that of the peritectoid and held there for some hours, the structure is found to be composed almost entirely of $\beta$, with a few particles of $\alpha$ remaining untransformed (Fig. 9-3$C$). Evidently the $\gamma$ transforms to $\beta$ with little or no composition change, and subsequently, the $\alpha$ is gradually dissolved by the $\beta$. The reaction to form $\beta$ begins at the $\alpha\gamma$ phase interface

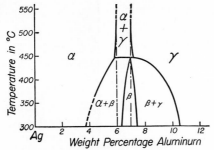

FIG. 9-2. Portion of the phase diagram of the system silver-aluminum.

and proceeds into the $\gamma$ phase at a rapid rate. This can be seen in the microstructure of the alloy in its partly transformed condition (Fig. 9-3$B$). Three phases ($\alpha$, $\beta$, and $\gamma$) are present in this microstructure. Since, according to the phase rule, there can be but two phases present at equilibrium in this type of situation, it is clear that the microstructure of Fig. 9-3$B$ represents a nonequilibrium state.

A hypoperitectoid alloy, 6% aluminum, quenched from above the peritectoid temperature, is found to be composed of a matrix of $\alpha$ with embedded particles of $\gamma$ (Fig. 9-4$A$). Again, with long heat treatment below the reaction temperature, the $\gamma$ is converted to $\beta$ and much of the $\alpha$ is consumed by solution in the $\beta$ (Fig. 9-4$B$).

## Limiting Cases of the Peritectoid

Rather common among phase diagrams of the alloy systems is the occurrence of a limiting case of the peritectoid in association with a limiting

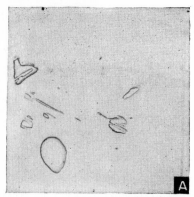

FIG. 9-3 (*A*). A near-peritectoid alloy, Ag + 7% Al, stabilized by long heat treatment above the peritectoid temperature and then quenched. "Islands" are the α phase embedded in a matrix of γ. Magnification 150.

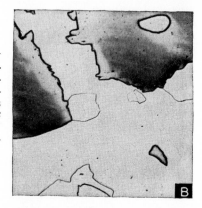

FIG. 9-3 (*B*). Same as *A*, partially transformed by cooling to a temperature slightly below that of the peritectoid and holding 20 min before quenching. Much of the light-colored γ has been transformed to dark-colored β without greatly affecting the "islands" of α, which are also light colored. Magnification 150.

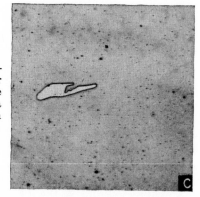

FIG. 9-3 (*C*). Same as *A*, fully transformed by cooling to and holding 2 hr at a temperature slightly below the peritectoid. Dark matrix is β; light area is residual α that has not yet been dissolved by the β. Magnification 150.

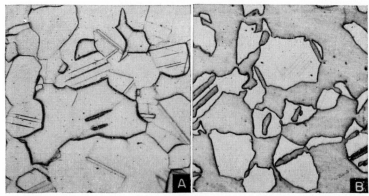

FIG. 9-4. (A) A hypoperitectoid alloy, Ag + 6% Al, stabilized by long heating at a temperature somewhat above that of the peritectoid, followed by quenching. The heavily outlined areas are $\gamma$; the $\alpha$ matrix is distinguished by exhibiting occasional twin bands across the grains. Magnification 150. (B) Same as A, fully transformed by holding 2 hr at a temperature slightly below the peritectoid, followed by quenching. The gray matrix is $\beta$; the lighter areas are residual $\alpha$. Magnification 150.

case of the eutectoid (see Fig. 9-5). Here the peritectoid and eutectoid compositions each coincide very nearly with one of the ends of their respective reaction isotherms. It is possible to tell that one of these re-actions is a peritectoid and the other a eutectoid by observing the relative positions of the two-phase fields. With peritectoid reaction two two-phase fields ($\beta + \beta'$ and $\beta' + \gamma$) lie below the reaction isotherm and one

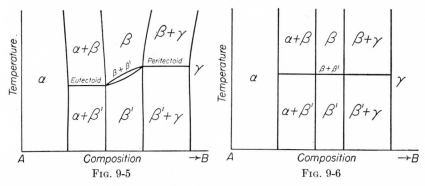

FIG. 9-5.     FIG. 9-6.

($\beta + \gamma$) above, while the reverse is true of the eutectoid. Occasionally, there is no perceptible temperature difference between the two reactions, and the connecting two-phase field ($\beta + \beta'$) reduces to a horizontal line (see Fig. 9-6). This is a limiting condition which probably is never actu-ally attained but which is so closely approached that the diagram cannot be drawn otherwise. Equilibria of this variety are sometimes associated with a change from an ordered to a disordered crystal; $\beta'$, for example, may be "ordered," while $\beta$ is "disordered."

CHAPTER 10

# BINARY SYNTECTIC SYSTEMS

Corresponding to the monotectic of the eutectic group, the *syntectic reaction* of the peritectic group consists, with rising temperature, in the decomposition of a solid phase into two immiscible liquids.

$$L_\mathrm{I} + L_\mathrm{II} \underset{\text{heating}}{\overset{\text{cooling}}{\rightleftharpoons}} \beta$$

This case is illustrated in Fig. 10-1. The termination in this diagram of the two-liquid field by an upper critical point is but one of several possibilities, as has been pointed out in Chap. 6. Likewise, the completion of the diagram with two eutectic reactions (dashed lines) is an arbitrary choice not involved in the syntectic itself.

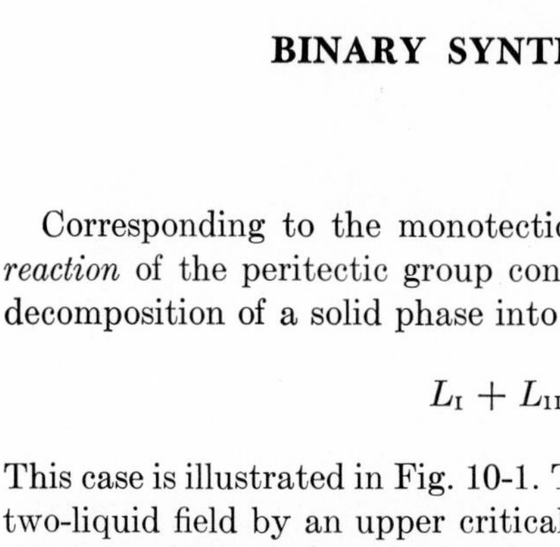

FIG. 10-1

No presently useful alloys occur within the range of any of the known syntectic reactions, and as a consequence, very little is known of their transformation behavior. It is easily seen from the phase diagram, however, that any molten alloy within the composition range of the syntectic line must separate into two liquids, presumably existing as separate layers in the container. Freezing should begin at the syntectic reaction isotherm, where $L_\mathrm{I}$ should react with $L_\mathrm{II}$ to form $\beta$. Ordinarily, the $\beta$ would form at the interface between the liquid layers and tend to interfere with further reaction to the extent that no substantial layer of $\beta$ would form except with very long time at the reaction temperature. Thus, it is to be expected that the "alloy" will freeze essentially in two independent parts with a minor reaction layer between. The $L_\mathrm{I}$ should freeze as a hypereutectic alloy of the $\alpha + \beta$ eutectic system, and the $L_\mathrm{II}$ as a hypoeutectic alloy of the $\beta + \gamma$ eutectic system. Only the relative proportions of the two liquids would change with composition from one end of the

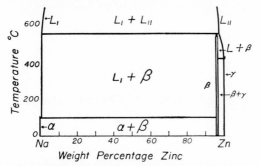

FIG. 10-2. Phase diagram of the system sodium-zinc.

syntectic line to the other; i.e., there would be nothing to distinguish the syntectic alloy from compositions on either side of it (except, of course, upon the attainment of true equilibrium).

Perhaps the best known example of the relatively rare syntectic is that found in the system sodium-zinc (Fig. 10-2). Here again, as was found to be the case in the peritectoid of the silver-aluminum system (Chap. 9) the composition of the low-temperature phase $\beta$ overlaps that of one of the high-temperature liquid phases $L_{II}$. Above the syntectic temperature an "alloy" composed of equal parts of sodium and zinc exists in the form of two liquid layers $L_I$ and $L_{II}$. Upon cooling, the zinc-rich liquid $L_{II}$ transforms immediately and quantitatively to solid $\beta$, leaving the sodium-rich layer $L_I$ unchanged until it finally freezes as a limiting eutectic just below the melting temperature of sodium. Since this change is effected virtually without the aid of diffusion, it is not delayed by interposing a layer of solid between

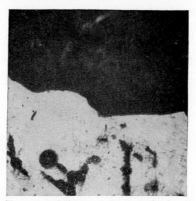

FIG. 10-3. Microstructure at the junction between the two layers in a cast "alloy" composed of equal parts of sodium and zinc. The lower layer is the $\beta$ (compound) phase; the upper layer is nearly pure sodium. Dark spots in the $\beta$ layer are inclusions of sodium. Magnification 75.

the two liquid layers. The resulting microstructure is composed of a one-phased layer of $\beta$ in line contact with a nearly one-phased layer of $\alpha$ (sodium) (Fig. 10-3).

# CHAPTER 11

# COMPLEX BINARY PHASE DIAGRAMS

The structural units from which binary phase diagrams of metal systems are built have now been considered. It will be instructive to see how these are combined to create some of the well-known diagrams of complex configuration. Before proceeding with this, however, it will be helpful to review the structural units that will be used.

One-phase equilibrium, being tervariant, may be represented by an area of any shape which fills the space not occupied by the two-phase regions. There must be an area for each phase existing in the system, although sometimes the area may be of vanishingly narrow width.

Two-phase equilibrium is bivariant and must be represented by a pair of conjugate bounding lines, every point on one boundary being connected

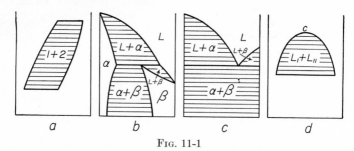

FIG. 11-1

by means of an (isothermal) tie-line with a conjugate point on the other boundary. Thus, it can be seen that the two-phase region must exist over a temperature range [except in the very rare limiting case where it exists as a horizontal line having no detectable area, (see Fig. 9-6)] and the two boundaries must, of course, span the same temperature range. Also, the boundaries must extend over ranges of composition. They are, therefore, never precisely straight and vertical, although they may very closely approach this condition. In its most general form the two-phase region appears as in Fig. 11-1a. Several of the ways in which the two-phase region may be drawn are illustrated in the other sketches of Fig. 11-1. Thus, in diagram b it is seen that the boundaries of the two-phase regions meet at points of congruent transformation (melting, in this example)

98

and that they terminate at separate points (without meeting) on a three-phase reaction isotherm. In the absence of information to the contrary, the $\alpha + \beta$ region of this diagram may be presumed to terminate at the absolute zero of temperature without closure of the boundary lines. Diagram $c$ represents a case where the one-phase $\alpha$ and $\beta$ regions are so narrow that the boundaries of the two-phase regions appear to be coincident with the composition lines of the pure components. The termination of a two-phase region in a critical point is illustrated in diagram $d$.

Three-phase equilibrium is univariant and is always represented by a horizontal line, the ends of which touch two one-phase regions and some intermediate point being in contact with a third one-phase region. Three two-phase regions terminate on every three-phase isotherm. Two classes of three-phase equilibrium have been distinguished: (1) the *eutectic type*, which may be represented thus:

$$2 \overbrace{\qquad\vee\qquad} 3$$

or thus:       $$1 \xrightarrow[\text{Heating}]{\text{Cooling}} 2+3,$$

and (2) the *peritectic type* which may be represented by:

$$1 \overbrace{\quad\wedge\quad}_{3} 2$$

or:       $$1+2 \xrightarrow[\text{Heating}]{\text{Cooling}} 3.$$

In each of these classes three specific combinations have been named:

Eutectic:    $L \rightleftharpoons \alpha + \beta$
Eutectoid:   $\gamma \rightleftharpoons \alpha + \beta$
Monotectic: $L_{\mathrm{I}} \rightleftharpoons \alpha + L_{\mathrm{II}}$

and       Peritectic:    $\alpha + L \rightleftharpoons \beta$
Peritectoid:   $\alpha + \gamma \rightleftharpoons \beta$
Syntectic:   $L_{\mathrm{I}} + L_{\mathrm{II}} \rightleftharpoons \beta$

It will be evident that other reactions of both classes should be possible. In none of the above cases has the gas phase been considered, nor has reaction wholly within the liquid state been mentioned. Fundamentally, there is no reason why any combination of three phases could not be substituted for those discussed, although the occurrence in nature of some combinations appears very improbable. There can, of course, be only one gas phase in any combination, because all gases are completely miscible. Three conjugate liquid and three conjugate solid phases are possible.

All conceivable combinations of the two types of three-phase equilibrium are listed in Tables 2 and 3.

Besides the six three-phase reactions that have been considered previously, there are three others that are definitely known to occur in systems involving metals; these reactions, which have no generally accepted names, are *l, e,* and *v* of the tables.

TABLE 2. EUTECTIC TYPES

| | | |
|---|---|---|
| a. | $L_I$ ⟩—G—⟨ $L_{II}$ | ⎫ |
| b. | $L_I$ ⟩—G—⟨ $\alpha$ | ⎬ Probably occur at low pressures or at high temperatures in metal systems |
| c. | $\alpha$ ⟩—G—⟨ $\beta$ | ⎪ |
| d. | $G$ ⟩—$L_I$—⟨ $L_{II}$ | ⎭ |
| e. | $G$ ⟩—$L_I$—⟨ $\alpha$ | Commonly occurs in gas-metal systems |
| f. | $L_{II}$ ⟩—$L_I$—⟨ $L_{III}$ | No case recorded among metal systems |
| g. | $\alpha$ ⟩—$L_I$—⟨ $L_{II}$ | Monotectic |
| h. | $\alpha$ ⟩—$L_I$—⟨ $\beta$ | Eutectic |
| i. | $G$ ⟩—$\alpha$—⟨ $L_I$ | Probably does not occur |
| j. | $G$ ⟩—$\alpha$—⟨ $\beta$ | Believed to occur in certain gas-metal systems |
| k. | $L_I$ ⟩—$\alpha$—⟨ $L_{II}$ | Probably does not occur |
| l. | $L_I$ ⟩—$\alpha$—⟨ $\beta$ | A rare case found in some metal systems |
| m. | $\beta$ ⟩—$\alpha$—⟨ $\gamma$ | Eutectoid |

Reaction *l*, which may be written

$$\alpha \underset{\text{heating}}{\overset{\text{cooling}}{\rightleftharpoons}} \beta + L$$

is thought to occur in copper-tin alloys between 38 and 58% tin (outside the range of commercial alloys). The melt freezes completely and then at a lower temperature (635°C) exhibits liquation.

Several gas-metal systems, of which the silver-oxygen system is the best known, undergo reaction *e*.

$$L \underset{\text{heating}}{\overset{\text{cooling}}{\rightleftharpoons}} \alpha + \text{gas}$$

The liquid freezes with simultaneous evolution of gas, which may appear as bubbles rising to the top of the melt or as cavities in the solidified alloy. If the gas is evolved under some pressure late in the freezing process, as happens with silver-oxygen, it may drive the last of the liquid out of the interstices between the solid dendrites with such force that liquid and gas are ejected from the surface of the nearly solid casting, causing behavior known as "spitting."

TABLE 3. PERITECTIC TYPES

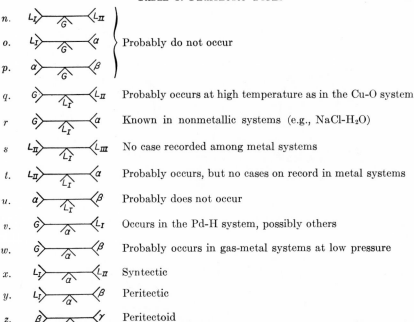

*n.* $L_I$ ／G ＼ $L_{II}$

*o.* $L_I$ ／G ＼ $\alpha$  } Probably do not occur

*p.* $\alpha$ ／G ＼ $\beta$

*q.* G ／$L_I$＼ $L_{II}$　　Probably occurs at high temperature as in the Cu-O system

*r* G ／$L_I$＼ $\alpha$　　Known in nonmetallic systems (e.g., NaCl-H$_2$O)

*s* $L_{II}$ ／$L_I$＼ $L_{III}$　　No case recorded among metal systems

*t.* $L_{II}$ ／$L_I$＼ $\alpha$　　Probably occurs, but no cases on record in metal systems

*u.* $\alpha$ ／$L_I$＼ $\beta$　　Probably does not occur

*v.* G ／$\alpha$＼ $L_I$　　Occurs in the Pd-H system, possibly others

*w.* G ／$\alpha$＼ $\beta$　　Probably occurs in gas-metal systems at low pressure

*x.* $L_I$ ／$\alpha$＼ $L_{II}$　　Syntectic

*y.* $L_I$ ／$\alpha$＼ $\beta$　　Peritectic

*z.* $\beta$ ／$\alpha$＼ $\gamma$　　Peritectoid

The reverse of this effect is observed with reaction *v*:

$$\text{Gas} + L \underset{\text{heating}}{\overset{\text{cooling}}{\rightleftharpoons}} \alpha$$

Solid palladium dissolves a considerable quantity of hydrogen, which, upon melting, "boils off" as gas bubbles.

### Assembly of Structural Units into a Phase Diagram

In the assembly of the structural units into a complete phase diagram the following principles should be observed:

1. One-phase regions may touch each other only at single points (i.e., points of congruent transformation), never along a boundary line.

2. Adjacent one-phase regions are separated from each other by two-phase regions involving the same two phases.

3. Three two-phase regions must originate upon every three-phase

isotherm; i.e., six boundary lines must radiate from each three-phase reaction horizontal.

4. Two three-phase isotherms may be connected by a two-phase region provided that there are two phases which are common to both of the three-phase equilibria.

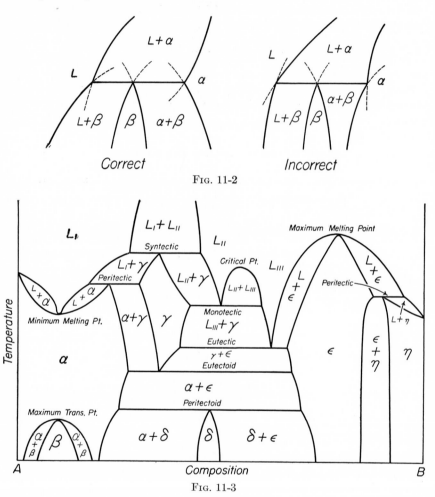

FIG. 11-2

FIG. 11-3

5. All boundaries of two-phase fields must project into two-phase fields where they join a three-phase isotherm, never into one-phase fields (see Fig. 11-2 and Appendix V).

The application of these rules in the construction of a complex diagram is illustrated by the imaginary phase diagram of Fig. 11-3, wherein all the structural units that have been discussed in previous chapters are included. One way of viewing complex diagrams is to consider that any binary diagram is made up of a characteristic group of univariant equilibria whose establishment fixes the identity and approximate location of all the bivariant equilibria. The univariant equilibria are, of course, the melting, boiling, and allotropic transformation points of the components of the system, including those of congruent intermediate phases, and the three-phase equilibria represented by isothermal reaction lines. To these may be added the critical points where two-phase equilibria terminate. Having located all the univariant equilibria on the phase diagram, there is usually only one possible arrangement of the two-phase regions, although the exact positions of the two-phase boundaries must be established from additional experimental data.

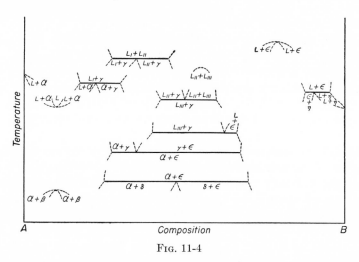

Fig. 11-4

Imagine, for example, that the melting points of the pure metals $A$ and $B$ (Fig. 11-3) and the points of congruent melting and transformation of the $\alpha$ and $\epsilon$ phases are known and that the horizontal reaction lines have all been inserted (Fig. 11-4). A two-phase region issues from each point of congruent transformation, and it can be named, because the two phases involved are the two phases that partake in the congruent transformation itself. Three two-phase regions issue from each horizontal reaction line, and these also can be named, because they constitute all possible pairs of the three phases involved in the three-phase reaction, i.e., in the reaction $\gamma \rightleftharpoons \alpha + \epsilon$ the two-phase pairs will be $\alpha + \gamma$, $\gamma + \epsilon$, and $\alpha + \epsilon$. If all the two-phase regions are thus correctly designated at their termini, it will be found that the two-phase boundaries can be drawn without ambiguity. Where the same pair of phases has been named twice, connecting boundary lines are drawn to close the two-phase region. If the same pair is named four times, as, for example, $L_I + \alpha$ or $L_3 + \epsilon$ in Fig. 11-4, it will be evident that two systems are involved and that the two-phase regions should be closed in such a way that no two-phase region crosses the composition of congruency. This applies also to the $\alpha + \beta$ equilibria in Fig. 11-4. If a two-phase pair is found to have no matching pair elsewhere on the diagram, it may be associated with a two-phase equilibrium that extends beyond the limits of the diagram. Thus, in Fig. 11-4, the regions $\alpha + \beta$, $\alpha + \delta$, $\delta + \epsilon$, and $\epsilon + \eta$ are terminated at the base line of the diagram and may be presumed to extend to the absolute zero of temperature without closure; region $L_I + L_{II}$ having no critical point may be presumed to terminate upon some three-phase reaction line at a higher temperature beyond the limits of the diagram. Care must, of course, be taken to ensure that the boundaries are so drawn that all projections extend into two-phase regions.

Although this manner of thinking about the synthesis of phase diagrams may seem artificial and, perhaps, too remote from questions of interpretation, it is commonly observed that a little practice in the construction of imaginary diagrams is very helpful in developing a "feel for phase diagrams." This feel is almost indispensable in the interpretation of diagrams taken at random from the literature, because instances of the misinterpretation of experimental data, resulting in an impossible construction of the diagram, are encountered from time to time and must be recognized if the user of the diagram is to make a sound judgment of the conclusions that may be derived therefrom. For practice purposes a set of problems of this kind is included at the end of this chapter. It is intended that the data given should be inserted on the coordinate frame of a diagram and that the diagram should then be completed with all regions properly labeled. A diagram having been completed, the list of rules given in the previous section should be checked to see that no errors have been made.

### Some Errors of Construction

Some fundamental errors of construction which should be recognized easily are presented in Fig. 11-5. Diagram $a$ is impossible because the three-phase equilibrium $L \rightleftharpoons \alpha + \beta$ is not isothermal (is not univariant) and therefore defies the phase rule. Similarly, the range of composition over which $\beta$ is represented in the $L + \alpha \rightleftharpoons \beta$ reaction of diagram $b$ is incompatible with univariant equilibrium. Four phases cannot partake in equilibrium (diagram $c$) except at a point of invariant equilibrium, the occurrence of which is highly improbable. The projection of two-phase boundaries must not extend into one-phase regions, as in diagram $d$, because this construction represents thermodynamic instability.

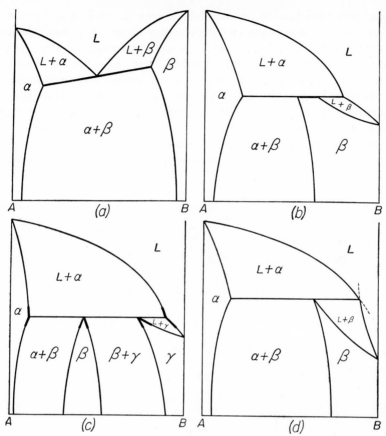

FIG. 11-5. Errors in phase-diagram construction.

## Interpretation of Complex Phase Diagrams

The prediction of the course of transformation and of the resulting alloy structure in systems of complex constitutional type differs from that in simple systems only in that a sequence of wholly or partially completed transformations may have to be considered. This matter has received some incidental attention in earlier chapters. It will be recalled, for example, that upon the completion of monotectic decomposition (Chap. 6), a liquid phase $L_{II}$ remains distributed in pockets throughout the solid $\alpha$ phase; upon subsequent cooling this liquid undergoes eutectic reaction while the $\alpha$ phase remains unchanged.

The principle of interpretation illustrated here may be stated as follows: With each successive phase change, as the temperature is altered, the alloy is broken up into constituent particles of different kinds, each kind having its own composition (or the reverse process in which two phases combine to form one may occur). Thenceforward each particle behaves in accordance with its composition, undergoing such changes as the phase diagram reports for that composition. Where the composition difference is continuous, as in a cored solid solution, then each tiny layer of substantially uniform composition should be regarded as capable of unique behavior in accordance with the constitutional demands to which it is subject.

Superimposed upon changes that accompany temperature rise or fall is a constant tendency to establish heterogeneous equilibrium among the

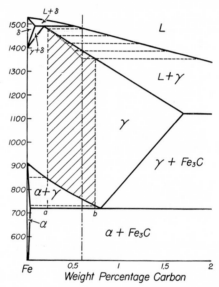

Fig. 11-6. Phase diagram of the metastable system Fe-C. Shaded zone represents the composition span in the cored 0.6% C alloy immediately after the completion of solidification and before eutectoid transformation.

different particles by the operation of diffusion. The higher the temperature and the slower the temperature change, the more important does this factor become.

No quantitative treatment of the relative influences of these factors is possible in any complex system at the present time. The best that can be done is to obtain a qualitative view of the process of successive transformation. An example will be used to illustrate. If an iron-carbon alloy, 0.6% carbon, is cast from the molten state and is allowed to cool slowly to room temperature, a series of changes will occur (see Fig. 11-6). Beginning at the liquidus temperature, a cored $\gamma$ phase will appear, which when fully solidified will possess a composition range of 0.3 to 0.75% carbon. Each layer of differing composition will reach the $\gamma$ boundary of the $\alpha + \gamma$ field at a different point (in composition and temperature) and will transform accordingly. The layer of composition $a$ will reject a large quantity of proeutectoid $\alpha$ and a little eutectoid pearlite, while the layer of composition $b$ will reject a little proeutectoid $\alpha$ and much eutectoid pearlite, intermediate layers behaving in intermediate fashion. Thus, the cored structure of the original $\gamma$ solid solution results in a segregation of the constituents of subsequent transformation.

This effect is most pronounced in the example of Fig. 11-7, which shows a banded structure resulting from dendritic segregation (coring) in a 0.3% carbon steel. Light-colored areas are those predominating in the $\alpha$ phase; darker areas are those composed of the pearlite constituent.

Obviously each combination of phase changes will produce a different result, making available a large variety of structures among the alloy

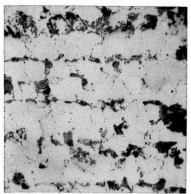

FIG. 11-7. Binary steel, 0.3% C, showing banding of the eutectoid constituent resulting from dendritic segregation. Magnification 100.

systems. None of these presents any particular difficulty in interpretation provided that the phase changes are considered one at a time and the influence of each change upon each of the constituents of the alloy is noted and carried forward to the next transformation.

Isothermal diffusion behavior in complex systems presents no new factors. An example is given in Fig. 11-8, where the structure produced by permitting blocks of copper and zinc to lie in contact for several days at 400°C is shown. It will be observed that there is a layer in the "alloy" corresponding to each *one-phase* region crossed by the horizontal line drawn across the phase diagram at 400°C. No layers corresponding to two-phase regions are formed, the two-phase regions being represented in the sample by the interfaces between the one-phase layers. It should be noted, however, that if diffusion had been permitted to continue until equilibrium was approximated, no more than two layers would persist in the sample. These could correspond to any two adjacent phases of the system, depending upon the gross composition of the sample. Indeed, at equilibrium, the structure of the diffusion sample would be given by applying the lever principle to the tie-line which crosses the gross composition at the temperature of diffusion. If the gross composition falls within a one-phase region, the "alloy" at equilibrium will, of course, be composed of but one phase.

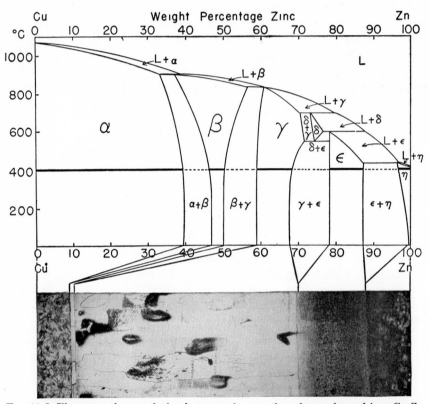

Fig. 11-8. Illustrates the correlation between the one-phase layers formed in a Cu-Zn diffusion couple and the Cu-Zn phase diagram. The horizontal line across the phase diagram at 400°C represents the sequence of states occurring in the diffusion couple. Oval spots in the γ layer of the diffusion couple are holes, resulting from the fact that Zn has diffused to the left more rapidly than Cu has diffused toward the right. Magnification 500.

The reason for the absence of two-phase layers in isothermal diffusion structures is to be found in the requirement that a composition gradient must exist for diffusion to occur. In a two-phase alloy, each phase tends toward a fixed equilibrium composition (as dictated by the appropriate tie-line) and no composition gradient can be maintained across a layer of the aggregate. The only composition difference that can be maintained is that between the two phases. Thus, diffusion can proceed only across the two-phase interface and not through the two-phase structure.

The $\gamma$ layer in the microstructure of the diffusion couple shown in Fig. 11-8 exhibits several dark oval spots; these are cavities.[1] They formed because there is a large difference between the diffusion velocities of zinc and of copper. The zinc diffuses faster, and the space vacated agglomerates into cavities.

## PRACTICE PROBLEMS

**1.** $A$ melts at 1000°, $B$ melts at 500°. There is a peritectic reaction: $\alpha$ (5% $B$) + $L$ (75% $B$) $\rightleftharpoons \beta$ (50% $B$) at 800° and another peritectic reaction: $\beta$ (55% $B$) + $L$ (90% $B$) $\rightleftharpoons \gamma$ (80% $B$) at 600°. Complete the diagram.

**2.** $A$ melts at 1000°; $B$ melts at 700°. There is a peritectic reaction: $\alpha$ (5% $B$) + $L$ (50% $B$) $\rightleftharpoons \beta$ (30% $B$) at 800°; a eutectic reaction: $L$ (80% $B$) $\rightleftharpoons \beta$ (60% $B$) + $\gamma$ (95% $B$) at 600°; and a eutectoid reaction: $\beta$ (50% $B$) $\rightleftharpoons \alpha$ (2% $B$) + $\gamma$ (97% $B$) at 400°. Complete the diagram.

**3.** $A$ melts at 800°; $B$ melts at 700°; the $\beta$ phase melts congruently at 1000°, its composition being 50% $B$. The following isothermal reactions occur:

Monotectic: $L_I$ (60% $B$) $\rightleftharpoons \beta$ (55% $B$) + $L_{II}$ (80% $B$), at 950°
Peritectic: $\beta$ (85% $B$) + $L_{II}$ (95% $B$) $\rightleftharpoons \sigma$ (52% $B$), at 800°
Eutectic: $L_I$ (20% $B$) $\rightleftharpoons \alpha$ (5% $B$) + $\beta$ (40% $B$), at 650°
Eutectoid: $\beta$ (45% $B$) $\rightleftharpoons \alpha$ (3% $B$) + $\sigma$ (90% $B$), at 500°
Peritectoid: $\alpha$ (2% $B$) + $\sigma$ (95% $B$) $\rightleftharpoons \gamma$ (50% $B$), at 300°

Complete the diagram.

**4.** Construct a phase diagram involving three peritectic reactions only.

**5.** Construct a phase diagram involving two monotectic and one eutectic reaction.

**6.** Construct a phase diagram involving two monotectic and one peritectic reaction.

**7.** Construct a phase diagram involving two peritectic, one eutectoid, and one peritectoid reaction.

**8.** Ascertain the order in which the several univariant temperatures would occur if one component having an allotropic transformation and another component having none are associated in a system involving ($a$) a peritectic and a peritectoid reaction or ($b$) a eutectic and a peritectoid reaction.

[1] For further details, see F. N. Rhines and B. J. Nelson, Structure of Copper-Zinc Alloys Oxidized at Elevated Temperatures, *Trans. Am. Inst. Mining Met. Engrs.*, **156**:171–194 (1944).

# CHAPTER 12

# TERNARY ISOMORPHOUS SYSTEMS

Ternary systems are those having three components. It is not possible to describe the composition of a ternary alloy with a single number or fraction, as was done with binary alloys, but the statement of two independent values is sufficient. Thus, the composition of an iron-chromium-nickel alloy may be described fully by stating that it contains 18% chromium and 8% nickel. There is no need to say that the iron content is 74%. But the requirement that two parameters must be stated to describe ternary composition means that two dimensions must be used to represent composition on a complete phase diagram.

The external variables that must be considered in ternary constitution are, then, temperature, pressure, composition $X$, and composition $Y$. To construct a complete diagram representing all these variables would require the use of a four-dimensional space. This being out of the question, it is customary to assume pressure constant (atmospheric pressure) and to construct a three-dimensional diagram representing, as variables, the temperature and two concentration parameters. In any application of the phase rule, therefore, it should be recalled that one degree of freedom has been exercised in the initial construction of the three-dimensional diagram by electing to draw it at 1 atmosphere of pressure.

Three-dimensional diagrams are usually so plotted that the composition is represented in the horizontal plane and the temperature vertically (see Fig. 12-4). Two-dimensional sections through this space model may be taken *horizontally* (isothermally) at various temperature levels or *vertically*. Since two-dimensional sections are more convenient than space models, their use is preferred. The selection and construction of such sections will be considered at length in this and subsequent chapters.

## The Gibbs Triangle

Because of its unique geometric characteristics, an equilateral triangle provides the simplest means for plotting ternary composition. Other types of axes that have been used from time to time will not be discussed in this book. On the *Gibbs triangle*, which is an equilateral triangle, the three pure component metals are represented at the corners, $A$, $B$, and $C$, Fig. 12-1. Binary composition is represented along the edges, i.e., the binary systems $AB$, $AC$, and $BC$. And ternary alloys are represented within the area of the triangle, such as at point $P$ in Fig. 12-1.

If lines are drawn through point $P$ parallel to each of the sides of the triangle, it will be found that these have produced three smaller equilateral triangles: $aaa$, $bbb$, and $ccc$. The sum of the lengths of the nine sides of these three triangles is equal to the sum of the lengths of the three sides of the major triangle $ABC$ within which they are inscribed, or the sum of the lengths of one side from each of the minor triangles is equal to the length of one side of the major triangle: $a + b + c = AB = AC = BC$. Also, the sum of the altitudes of the minor triangles is equal to the altitude of the major triangle: $a' + b' + c' = AX$.

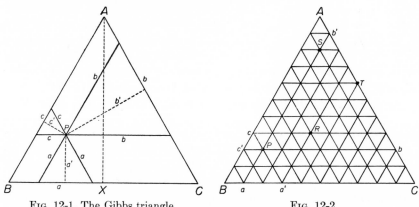

FIG. 12-1. The Gibbs triangle.     FIG. 12-2

If one side of the Gibbs triangle be divided into 100 equal parts, representing 100% on the binary composition scale, it will be found that the same units can be used to measure the composition at point $P$. Let the length $a$ represent the percentage of $A$ in $P$, the length $b$ the percentage of $B$, and the length $c$ the percentage of $C$. Since these lengths total the same as one side of the Gibbs triangle and together they must equal 100%, it is evident that 1% has the same length, whether measured along an edge of the diagram or along any inscribed line parallel to an edge. A similar result could be obtained by using altitudes, but this is less convenient. It should be noted that in either case, the percentage of $A$ is measured upon the side of $P$ away from the $A$ corner and similarly with $B$ and $C$.

For convenience in reading composition, an equilateral triangle may be ruled with lines parallel to the sides (Fig. 12-2). Composition may then be read directly, for example, $P = 20\% A + 70\% B + 10\% C$. At point $P$ the percentage of $A$ is represented by the line $Pa$ (or equivalently $Pa'$), which is 20 units long; the percentage of $B$ by the line $Pb$ (or $Pb'$), 70 units long; and the percentage of $C$ by the line $Pc$ (or $Pc'$), 10 units long. Other examples shown in Fig. 12-2 are $R = 30\% \ A + 40\% \ B + 30\% \ C$, $S = 80\% \ A + 10\% \ B + 10\% \ C$, and $T = 60\% \ A + 0\% \ B + 40\% \ C$.

## Tie-lines

It can be shown that if any two ternary alloys are mixed together, the composition of the mixture will lie on a straight line joining the original two compositions. This is true regardless of the proportions of the two alloys in the mixture. Conversely, if an alloy decomposes into two fractions of differing composition, the compositions of the two portions will lie on opposite ends of a straight line passing through the original composition point. The truth of this statement may be tested by reference to the example illustrated in Fig. 12-3. Points $S$ and $L$ represent two ternary alloys of respective composition: $20\% \ A + 70\% \ B + 10\% \ C$ and $40\% \ A + 30\% \ B + 30\% \ C$. Suppose that one part of $S$ is mixed with three parts of $L$ and the mixture is analyzed. The analytical result must, of course, be

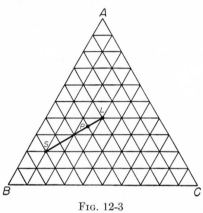

FIG. 12-3

$$0.25 \times 20\% \ A + 0.75 \times 40\% \ A = 35\% \ A$$
$$0.25 \times 70\% \ B + 0.75 \times 30\% \ B = 40\% \ B$$
$$0.25 \times 10\% \ C + 0.75 \times 30\% \ C = 25\% \ C$$

As can be seen by inspection of Fig. 12-3, this composition lies at $P$, which is a point on the straight line connecting $S$ and $L$. No matter what compositions had been chosen or in what proportions they had been mixed, the total composition would have occurred on the line joining the two original compositions.

It will be evident that the line $SL$ has the characteristics of a *tie-line;* it is both isobaric and isothermal, because it lies in the composition plane which is drawn perpendicular to the temperature axis and corresponds to the case of constant atmospheric pressure (i.e., would be drawn perpendicular to the pressure axis if a fourth dimension were available). The lever principle is applicable to this line as is indicated in the above demonstration. Hence, the line $SL$ might represent the condition of an alloy of composition $P$ which is partially frozen, at the temperature under consideration, and consists of 25% solid of composition $S$ and 75% liquid of composition $L$:

$$\%S = \frac{PL}{SL} \times 100 \qquad \text{and} \qquad \%L = \frac{SP}{SL} \times 100$$

### The Space Diagram

A temperature-composition (TXY) diagram of an isomorphous system appears in Fig. 12-4. The composition plane forms the base of the figure, and temperature is measured vertically. Here, the liquidus and solidus

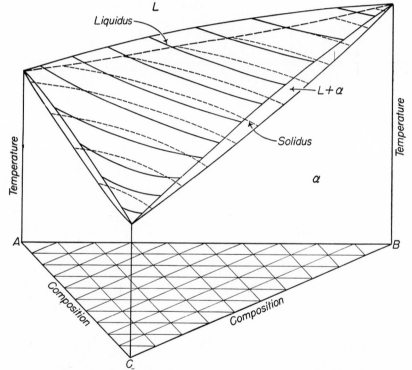

Fig. 12-4. Temperature-composition space diagram of a ternary isomorphous system.

become surfaces bounding the $L + \alpha$ space. Above the liquidus, all alloys are fully molten; below the solidus, all are completely solid. As in binary systems, the two-phase region $L + \alpha$ is composed of tie-lines joining conjugate liquid and solid phases. In the ternary system, however, the tie-lines are not confined to a two-dimensional area but occur as a bundle of lines of varying direction, but all horizontal (isothermal), filling the three-dimensional two-phase space.

## Isothermal Sections

The location of the tie-lines can be visualized more easily by reference to isothermal (horizontal) sections cut through the temperature-composition diagram at a series of temperature levels. The three isotherms presented in Fig. 12-6 are taken at the temperatures designated $T_1$, $T_2$, and $T_3$ in Fig. 12-5. It will be seen that the first tie-line on each edge of the

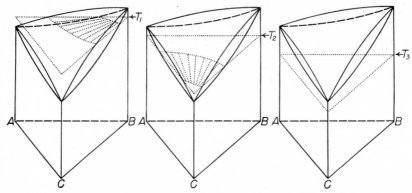

FIG. 12-5. Development of isotherms shown in Fig. 12-6.

$L + \alpha$ region is the bounding line of the figure; it is, in other words, the binary tie-line at the temperature designated. The directions of tie-lines lying within the figure vary "fanwise" so that there is a gradual transition from the direction of one bounding tie-line to that of the other. *No two tie-lines at the same temperature may ever cross.* Beyond this nothing

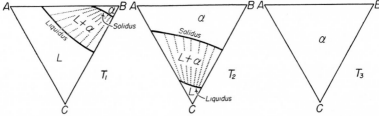

FIG. 12-6. Isotherms through the ternary isomorphous phase diagram, as derived from Fig. 12-5.

can be said of their direction, except, of course, that they must run from liquidus to solidus. Other than those on the edges of the diagram, none points toward a corner of the diagram unless by mere coincidence. It is necessary, therefore, to determine the position and direction of the tie-lines by experiment and to indicate them on the ternary phase diagram. Few of the published ternary diagrams show them because of a lack of sufficient data. There are a few cases, however, in which more intensive studies have been conducted and the tie-lines are recorded.

Isothermal sections such as those shown in Fig. 12-6 generally provide the most satisfactory means for recording ternary equilibria in two dimensions. In them the various structural configurations assume their simplest forms. The method becomes cumbersome, to be sure, when many isotherms are required to delineate the structure of the space diagram, but no alternate means has been devised to overcome this difficulty. Orthographic sketches of the space model are convenient for showing the general plan of the ternary diagram in a qualitative way, but they are not well adapted to the recording of quantitative data.

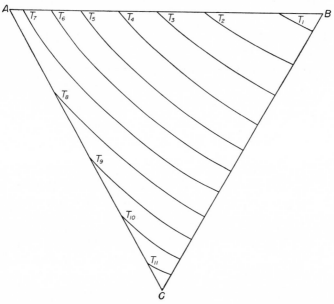

FIG. 12-7. Liquidus projection of the diagram shown in Fig. 12-4.

It is possible in certain cases to telescope a group of isotherms into a single diagram. For example, the isothermal lines which were used to delineate the liquidus surface in Fig. 12-4 may be projected onto a plane, such as the base of the diagram, giving the *liquidus projection* presented in Fig. 12-7. Each line is derived from a separate isotherm, and its temperature should therefore be indicated upon the line. In like manner the solidus (and other surfaces in more complex diagrams) may be presented as projections. Except in very simple systems, this scheme achieves only a minor condensation of the data, and it suffers from the handicap that tie-lines and other relationships existing among the parts of the diagram cannot be presented.

## Vertical Sections (Isopleths)

Because of their seeming resemblance to binary diagrams, vertical sections, also known as *isopleths,* have been widely used. The sections selected are usually (1) those radiating from one corner of the space diagram and, therefore, representing a fixed ratio of two of the components

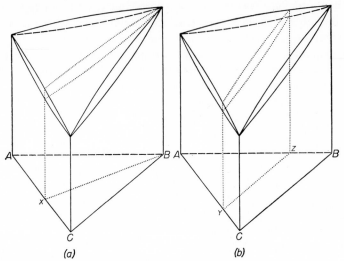

FIG. 12-8. Development of the isopleths shown in Fig. 12-9.

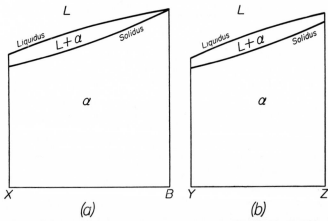

FIG. 12-9. Isopleths through an isomorphous system, derived from Fig. 12-8.

(Figs. 12-8a and 12-9a) or (2) those parallel to one side of the space diagram, representing a constant fraction of one of the components (Figs. 12-8b and 12-9b). It will be observed that the $L + \alpha$ region is open at its ends except where it terminates on the $B$ component. From vertical sec-

tions the liquidus and solidus temperatures for any of the alloys represented can be read. It is not possible, however, to represent tie-lines on
these sections, because no tie-line can lie in either section except by coincidence. In general, the tie-lines pass through the $L + \alpha$ region at an
angle to the plane of the section. Consequently, it is not possible to
record equilibria within the $L + \alpha$ region by the use of vertical sections.
Sometimes this is a matter of little consequence, when, for example, the
section lies so close to one side of the space diagram that the tie-lines may
be presumed to lie *approximately* in the section. For this reason, vertical
sections continue to be of some use, though they are gradually being displaced by isothermal sections in the literature of alloy constitution.
Certain vertical sections in complex diagrams (the quasi-binary sections)
do contain all tie-lines and will continue to be used; these are discussed in
Chap. 17.

### Application of the Phase Rule

Within the one-phase spaces of the ternary diagram, the $L$ and $\alpha$
regions of Fig. 12-4, the equilibria are quadrivariant; i.e., there are four
degrees of freedom.

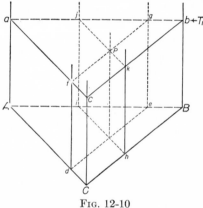

$$P + F = C + 2$$
$$1 + 4 = 3 + 2$$

These are pressure, temperature,
concentration $X$, and concentration
$Y$. The pressure was chosen when
the diagram was taken at 1 atmosphere of pressure. It remains to verify the correctness of representing
the one-phase regions by volumes
in the space diagram by ascertain-
ing if an independent choice of temperature and two concentration val-

Fig. 12-10

ues will describe a fixed equilibrium. Consider the diagram in Fig. 12-10;
this figure represents a portion of the $\alpha$ field of Fig. 12-4, the liquidus and
solidus having been removed. Let it be required that in addition to
adjusting the pressure at 1 atmosphere, the temperature is to be estab-
ilshed at $T_1$ and the composition at $20\%$ $A$ and $40\%$ $B$. A horizontal
plane $abc$, representing a choice of temperature, is passed through the
diagram at $T_1$. Next, one condition of composition is selected by erecting
a vertical plane $defg$ parallel to the $BC$ side and one-fifth of the distance
from this side to the $A$ corner. This plane intersects the $T_1$ isotherm along
the line $fg$. At this juncture it can be seen that the equilibrium is not yet

fully described, because any one of many alloys having 20% $A$ can exist as the $\alpha$ phase at 1 atmosphere of pressure and at temperature $T_1$. If a second vertical plane $hijk$, parallel to the $AC$ side of the diagram and two-fifths of the distance from this side to the $B$ corner, is erected, it will intersect the $T_1$ isotherm along the line $jk$ and will cross the line $fg$ at $P$. Thus, a unique point (point $P$) has been designated by the exercise of four degrees of freedom. It will be evident from the argument that has been followed that the phase rule in no way limits the shape of the one-phase regions.

When two phases coexist at equilibrium in a ternary system, there can be only three degrees of freedom:

$$P + F = C + 2$$
$$2 + 3 = 3 + 2$$

The system is tervariant. Pressure being fixed, there remain but two degrees of freedom, which can be temperature and one concentration value or two concentration values. Both of these possibilities will be explored. First, the choice of temperature and one concentration value will be examined.

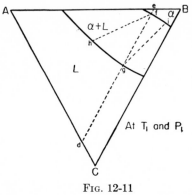

At $T_1$ and $P_1$

Fig. 12-11

Let a temperature be selected such that two-phase equilibrium may occur, for example, $T_1$ in Figs. 12-5 and 12-6. This isotherm is reproduced in Fig. 12-11. Two degrees of freedom have now been exercised, namely, pressure and temperature; a *single statement* concerning the composition of one of the phases should complete the description of a specific equilibrium. Therefore, let it be specified that the liquid phase shall contain 15% $A$. All alloys containing 15% $A$ are represented on the isotherm by the dashed line $de$. This line includes, however, only one composition of liquid which can be in equilibrium with the solid phase, namely, the composition $g$, where the dashed line crosses the liquidus. Thus, the composition of the liquid phase is described completely, and the composition of the conjugate solid is identified by the corresponding tie-line at point $i$ on the solidus. Had it been specified, instead, that the $\alpha$ phase should contain 15% $A$, then the composition of the solid would have been given at point $f$ and that of the conjugate liquid at $h$.

Two statements concerning the composition should serve also to identify the temperature and composition of both phases. This case is illustrated in Fig. 12-12. It has been specified that the liquid phase shall contain 5% $B$ and 10% $C$. Two vertical planes have been erected, one at a constant $B$ content of 5% and the other at a constant $C$ content of 10%. These intersect in the vertical line $L_1P$ which intersects the liquidus at $L_1$. This point on the liquidus lies in the isothermal section $abcd$ and is associated with a specific composition of the $\alpha$ phase at $\alpha_1$ by means of the tie-line $L_1\alpha_1$. Thus, the composition of the liquid phase, the composition of the solid phase, and the temperature of the equilibrium have all been established. Obviously, the same result could have been obtained by designating two concentration parameters for the solid phase or one each for the liquid and solid phases.

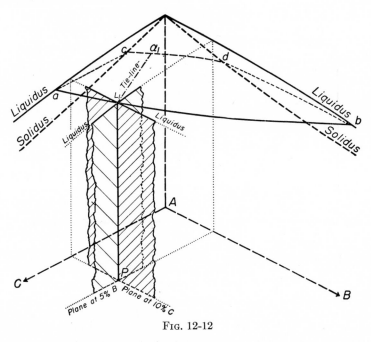

FIG. 12-12

From these demonstrations it will be apparent that the only construction of a two-phase region consistent with the phase rule is one in which there are two conjugate bounding surfaces existing within the same temperature range and upon which every point of one is connected by a horizontal tie-line with a unique point on the other.

### Equilibrium Freezing of Solid-solution Alloys

The course of equilibrium freezing of a typical alloy of a ternary isomorphous system may be followed by reference to Fig. 12-13. Beginning at a temperature within the liquid field and cooling at a rate that is "infinitely" slow, to permit the maintenance of equilibrium, the alloy of composition $X$ will begin to freeze when the liquidus surface is reached at temperature $T_1$. Here the liquid $L_1$ is in equilibrium with crystals of composition $\alpha_1$. As the solid grows, the liquid composition will change in a direction away from the $\alpha_1$ composition, following a curved path down the liquidus surface $L_1L_2L_3L_4$. At the same time the $\alpha$ composition approaches the gross composition $X$ along curved path $\alpha_1\alpha_2\alpha_3\alpha_4$. The tie-line, at each temperature, passes through the $X$ composition, joining the

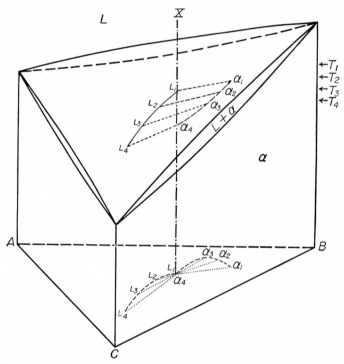

Fig. 12-13. Path of composition change of the liquid $L$ and solid $\alpha$ phase during the freezing of a solid-solution alloy.

conjugate compositions of $L$ and $\alpha$. Hence the tie-line lies in a different direction at each successive temperature.

Inscribed upon the base of the diagram of Fig. 12-13 is a projection of the tie-lines and the paths of composition variation of the liquid and solid phases. The liquidus path starts at the $X$ composition and ends at $L_4$, while the solidus path starts at $\alpha_1$ and ends at $X$. All tie-lines, $\alpha_1L_1$, $\alpha_2L_2$, $\alpha_3L_3$, and $\alpha_4L_4$, pass through $X$. The same information is presented in somewhat more realistic fashion in Fig. 12-14, where the complete isotherm for each of the temperatures from $T_1$ to $T_4$ is given. In these drawings it becomes apparent that the curved paths of the solid and liquid compositions result from the turning of the tie-lines to conform with the liquidus and solidus isotherms at successively lower temperatures.

## Natural Freezing of a Solid-solution Alloy

In natural freezing, the ternary solid solution departs from the course of "equilibrium" freezing in much the same way and for the same reason as does the binary solid-solution alloy. Diffusion fails to maintain the

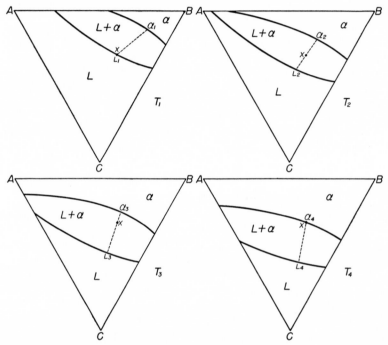

FIG. 12-14. Illustrating the progress of the "equilibrium freezing" of a ternary isomorphous alloy.

equilibrium compositions, so that a cored solid solution is produced and the temperature range over which freezing takes place is increased. This process is illustrated in Figs. 12-15 and 12-16, where the tie-line connecting the average solid and average liquid compositions is dashed. This line must, of course, pass through the gross composition $X$. Because both the solid and liquid phases present at the interface where freezing is in progress differ in their compositions from the corresponding averages, the solid tie-lines (solid lines in Fig. 12-16) representing "equilibria" at the front of freezing do not in general pass through the $X$ composition except at the start of the process. The average $\alpha$ composition must, obviously, change more gradually than does the $\alpha$ in contact with liquid, and the average composition of the liquid phase must similarly lag behind the composition change at the freezing front. Thus, the "average" tie-line

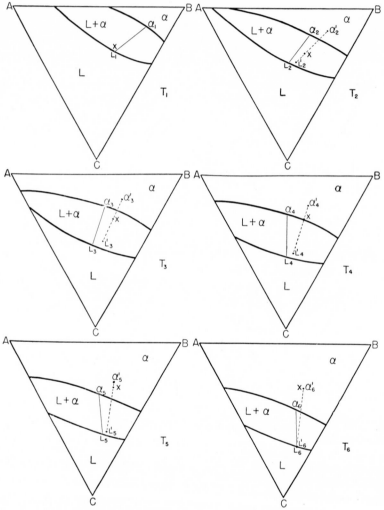

FIG. 12-15. Illustrating the progress of natural freezing of a ternary isomorphous alloy.

turns less with each step in cooling, shows an excess of liquid over the equilibrium quantity at each temperature, and proceeds to lower temperatures than are possible under equilibrium conditions. When the "average" composition of $\alpha$ coincides with $X$, the alloy is frozen. The coring produced in this way is complicated by the curved path of freezing on the solidus, but this complication is not apparent in the microstructure, which is indistinguishable from that of binary coring (Fig. 3-9*b*).

A homogenizing heat treatment may be used to reduce composition differences as with binary alloys. The selection of a suitable heat-treating temperature involves the same considerations as those cited for the simpler case. Liquation, too, may result from overheating in a manner analogous to liquation in binary alloys, being complicated only by the curved path of liquid composition that must be followed during melting

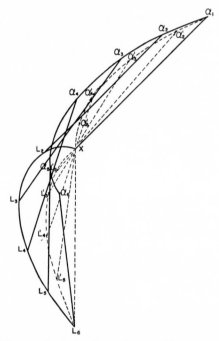

Figure 12-16. Tie-lines from the six isotherms of Fig. 12-15 superimposed.

and the curved path of solid composition followed during subsequent freezing. Changes of physical properties resulting from coring and subsequent heat treatment are likewise similar in kind.

## Maxima and Minima

Maxima and minima in ternary isomorphous systems are of two kinds, namely, (1) those occurring at a binary face of the diagram (Fig. 12-17) and (2) those occurring at a ternary composition (Fig. 12-18). In either case the liquidus and solidus meet at the maximum or minimum point. Where maxima or minima occur both at binary and at ternary composi-

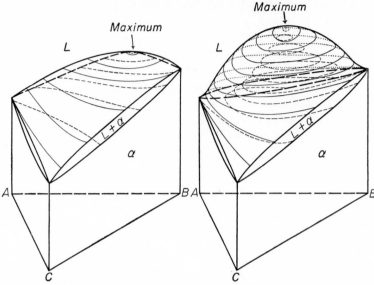

Fig. 12-17. Isomorphous ternary system with a maximum melting point in the binary system $AB$.

Fig. 12-18. Isomorphous ternary system with a maximum melting point in ternary space.

tions, the liquidus and solidus meet at both points, but not between the points. As with binary systems the alloys of maximum or minimum melting are congruently melting and may, if desired, be treated as components.

## PRACTICE PROBLEMS

**1.** Sketch the solidus projection corresponding to the liquidus projection of Fig. 12-7.

**2.** How should the shapes of the composition paths of the $\alpha$ and $L$ phases during freezing (Fig. 12-13) vary as the gross composition $X$ approaches the $BC$ side of the diagram, the $AB$ side, the $AC$ side, the $C$ corner, the $B$ corner, and the $A$ corner?

**3.** Draw isothermal sections corresponding to the temperatures at which liquidus contours are shown in Fig. 12-17; the same for Fig. 12-18.

**4.** Draw isopleths through the space diagram of Fig. 12-17 as follows: (a) from $C$ through the maximum in the binary system $AB$, (b) from 50% $B$ + 50% $C$ through the maximum in $AB$, (c) from $C$ to 50% $A$ + 50% $B$, (d) along the line of constant $B$ content 50%; using the space diagram of Fig. 12-18, draw isopleths (e) from $C$ through the temperature maximum to the $AB$ side of the figure, (f) from $C$ to the $AB$ side, passing slightly to the right of the maximum.

# CHAPTER 13

# TERNARY THREE-PHASE EQUILIBRIUM

Three-phase equilibrium in ternary systems occurs over a *temperature range* and not, as in binary systems, at a single temperature. It is bivariant:

$$P + F = C + 2$$
$$3 + 2 = 3 + 2$$

After the pressure has been established, only the temperature, or one concentration parameter, may be selected in order to fix the conditions of equilibrium. The representation of three-phase equilibrium on a phase diagram requires the use of a structural unit which will designate, at any given temperature, the fixed compositions of three conjugate phases. Such a structural unit is found in the tie-triangle.

## Tie-triangles

If any three alloys of a ternary system are mixed, the composition of the mixture will lie within the triangle produced by connecting the three original composition points with straight lines (Fig. 13-1). For example, take three compositions $R$, $S$, and $L$ as follows:

$$R = 20\% \ A + 70\% \ B + 10\% \ C$$
$$S = 40\% \ A + 40\% \ B + 20\% \ C$$
$$L = 10\% \ A + 30\% \ B + 60\% \ C$$

and mix two parts of $R$ with three parts of $S$ and five parts of $L$:

$$0.2 \times 20 + 0.3 \times 40 + 0.5 \times 10 = 21\% \ A$$
$$0.2 \times 70 + 0.3 \times 40 + 0.5 \times 30 = 41\% \ B$$
$$0.2 \times 10 + 0.3 \times 20 + 0.5 \times 60 = 38\% \ C$$

The total composition ($21\% \ A + 41\% \ B + 38\% \ C$) is seen to lie *within* the triangle $RSL$ at point $P$, Fig. 13-1. No matter what proportions of the three alloys had been taken, the same would have been true.

Once more the lever principle can be applied. This time the lever is the weightless plane triangle $RSL$ supported on a point fulcrum at $P$. If 20%

of the alloy $R$ is placed upon point $R$, 30% of alloy $S$ upon point $S$, and 50% of alloy $L$ upon point $L$, then the lever plane will balance exactly. For purposes of calculation it is convenient to resolve the planar lever into two linear levers such as $SPO$ and $ROL$, Fig. 13-1, by drawing a straight line from any corner of the triangle through point $P$ to its intersection with the opposite side. The quantity of the $S$ alloy in the mixture $P$ is then

$$\%S = \frac{PO}{SO} \times 100 \qquad \text{and} \qquad \%O = \frac{SP}{SO} \times 100$$

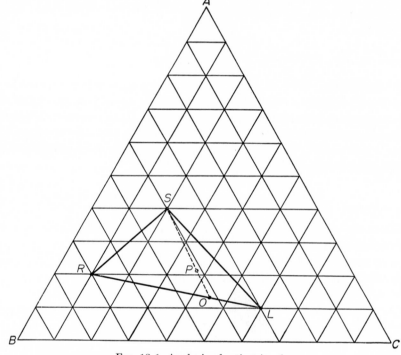

FIG. 13-1. Analysis of a tie-triangle.

Composition $O$ represents the mixture of alloys $R$ and $L$, so that

$$\%R = \frac{OL}{RL}\frac{SP}{SO} \times 100 \qquad \text{and} \qquad \%L = \frac{RO}{RL}\frac{SP}{SO} \times 100$$

By measuring the line $SPO$, it is found that the segment $SP$ is 18.2 mm long and the segment $PO$ is 7.8 mm long. Thus, the mixture $P$ contains

$$\%S = \frac{7.8}{26} \times 100 = 30\%$$

On the line $ROL$, segment $RO$ is 32.85 mm long and segment $OL$ is 13.15 mm long, whence

$$\%R = \frac{13.15}{46} \times \frac{18.2}{26} \times 100 = 20\%$$

$$\%L = \frac{32.85}{46} \times \frac{18.2}{26} \times 100 = 50\%$$

The triangle $RSL$ may be employed as a tie-triangle connecting three phases which associate to form an intermediate gross composition or phase or, conversely, three phases into which $P$ decomposes. There are then two kinds of "tie-elements" that appear in ternary diagrams, namely, the *tie-line* and the *tie-triangle*.

## Three-phase Equilibrium in the Space Model

A simple example of three-phase equilibrium in a ternary system is shown in Fig. 13-2. One of the binary systems involved in this ternary system is isomorphous; the other two are of the eutectic type. There are three one-phase regions: $L$, $\alpha$, and $\beta$; three two-phase regions: $L + \alpha$, $L + \beta$, and $\alpha + \beta$; and one three-phase region: $L + \alpha + \beta$. These are shown individually in the "exploded model," Fig. 13-3. Edges of regions that meet to form a common line in the assembled diagram are labeled with identical numbers in the exploded model.

Each of the two-phase regions is like a piece cut out of the $L + \alpha$ region of the isomorphous diagram discussed in the preceding chapter. It has two bounding surfaces which are connected at every point by tie-lines joining the conjugate phases. Thus, the $L + \alpha$ region of Fig. 13-3 exhibits a liquidus surface on top and a solidus surface beneath (dotted outline); its form can be seen in the matching top face of the $\alpha$ region shown just below. Tie-lines join every point on the liquidus with a corresponding point on the solidus. The other exposed surfaces of the $L + \alpha$ region are portions of the confining walls of the space diagram. On its underside the $L + \alpha$ region has a "ruled surface" (dotted outline) formed by the tie-lines (dotted) that join the edge of the liquidus 4 with the edge of the solidus 3. This surface is identical with one of the top surfaces of the $L + \alpha + \beta$ region 1-2-3-4. The $L + \beta$ region is entirely similar and its undersurface is identical with the other top surface of the $L + \alpha + \beta$ region 1-2-4-5. Beneath the three-phase region is the field of $\alpha + \beta$, which is bounded by two solvus surfaces that are connected at all points by tie-lines joining the conjugate $\alpha$ and $\beta$ phases. The top surface of the $\alpha + \beta$ region is generated by tie-lines connecting the edges 3 and 5. This surface is identical with the bottom surface of the $L + \alpha + \beta$ field, 1-2-3-5, not visible in the drawing (dotted tie-lines).

From this it can be seen that the three-phase region $L + \alpha + \beta$ is enclosed by three surfaces, each composed of tie-lines. At any one temperature, the tie-lines on the three surfaces meet to form a triangle, which is a tie-triangle. Just as each two-phase field is composed of a "bundle of tie-lines," the three-phase field is composed of a "stack of tie-triangles" (see Fig. 13-4). At each end of the field the last triangle is closed to a single line (the binary eutectic line), so that the region has only three surfaces, all of which meet at sharp edges on all sides. Although generated by horizontal straight lines (tie-lines), the three bounding surfaces are not

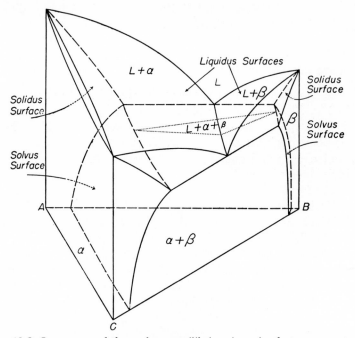

Fig. 13-2. Occurrence of three-phase equilibrium in a simple ternary system.

plane, because the tie-lines turn as they descend in temperature, producing curved ruled surfaces. The vertices of the tie-triangles lie on three conjugate lines labeled $\alpha$, $\beta$, and $L$ in Fig. 13-4. By reference to Figs. 13-2 and 13-3 it can be seen that the respective one-phase fields terminate on these lines. Thus, at each temperature level, three-phase equilibrium is represented by a tie-triangle the corners of which touch the three one-phase regions at composition points that are unique and the sides of which are in contact, respectively, with the three two-phase fields.

These details are perhaps best represented by the use of isothermal sections, Fig. 13-6, where the five isotherms are taken at the temperatures designated $T_1$ to $T_5$ in Fig. 13-5. Returning to the requirements of the

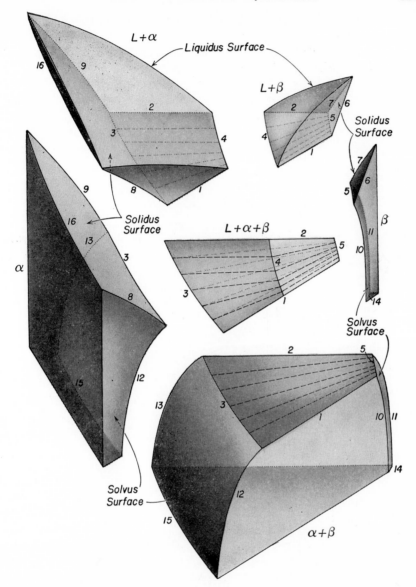

FIG. 13-3. Exploded model of the diagram of Fig. 13-2.

phase rule, with which this chapter was opened, it is now a simple matter to demonstrate that the one degree of freedom remaining after the establishment of a fixed pressure is sufficient to complete the definition of all variables in three-phase equilibrium. If a specific temperature is chosen, say $T_2$ in Fig. 13-6, then the compositions of all of the three conjugate

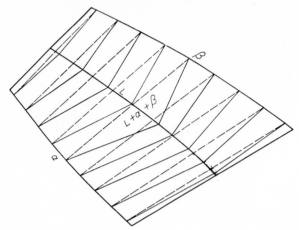

FIG. 13-4. A region of three-phase equilibrium in temperature-composition space.

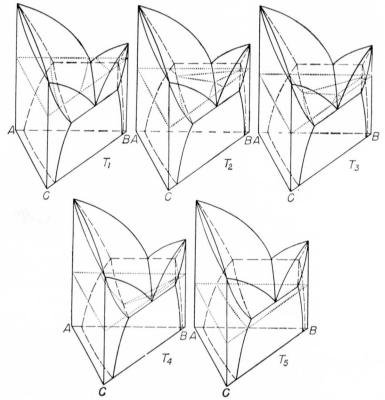

FIG. 13-5. Development of isotherms shown in Fig. 13-6.

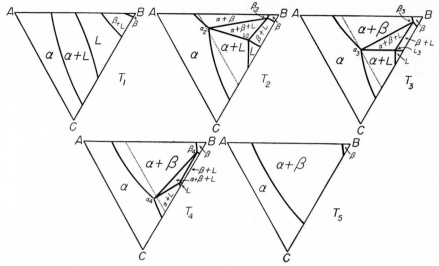

FIG. 13-6. Isotherms through the space diagram of Fig. 13-2.

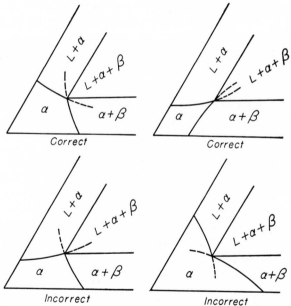

FIG. 13-7. Permissible and forbidden arrangements of the boundaries of the one-phase field.

phases are fixed, respectively, at $\alpha_2$, $\beta_2$, and $L_2$. Or similarly, if a concentration parameter is defined by requiring that the $\alpha$ phase in equilibrium with $\beta$ and $L$ shall contain 35% $B$ (indicated by the dotted line in isotherms $T_2$, $T_3$, and $T_4$), there will be only one isotherm ($T_3$) in which this condition is realized. The temperature and the compositions of the conjugate $\beta_3$ and $L_3$ phases are fixed by establishing one concentration parameter of $\alpha$.

The principle of the minimum free energy leads to a helpful limitation on the construction of the three-phase region, illustrated in Fig. 13-7. Boundaries between the one- and two-phase fields, where they meet the three-phase field, may either *both* project into different two-phase fields or *both* project into the three-phase field, but it is *not correct* for one of a pair of boundaries to project into a two-phase field and the other into the three-phase field or for either or both to project into the one-phase field.

### Freezing of an Alloy

Before considering the course of natural freezing of an alloy involving ternary three-phase equilibrium, it will be instructive to follow the corresponding path of "equilibrium phase change." This can be done most satisfactorily by reference to a series of isotherms (Fig. 13-8). The circled cross designates in each isotherm the gross composition $X$ of the alloy under observation. At temperature $T_1$ (the first isotherm) the $X$ composition lies in the liquid field and the alloy is fully molten.

At $T_2$, the next lower temperature represented, freezing is just beginning. Composition $X$ lies upon the liquidus and is joined with the first solid to appear by the tie-line $L_2\alpha_2$.

At $T_3$ a substantial quantity of $\alpha$ is present:

$$\%\alpha_3 = \frac{XL_3}{\alpha_3 L_3} \times 100 \approx 20\%$$

and
$$\%L_3 = \frac{\alpha_3 X}{\alpha_3 L_3} \times 100 \approx 80\%$$

At $T_4$, as the gross composition passes into the $L + \alpha + \beta$ field, the first particles of the $\beta$ phase begin to appear; the $\alpha$ and $\beta$ now crystallize simultaneously from the liquid.

At $T_5$ all three phases are present in substantial quantity:

$$\%\alpha_5 = \frac{XS_5}{\alpha_5 S_5} \times 100 \approx 60\%$$

$$\%\beta_5 = \frac{L_5 S_5}{L_5 \beta_5} \frac{\alpha_5 X}{\alpha_5 S_5} \times 100 \approx 20\%$$

and
$$\%L_5 = \frac{S_5 \beta_5}{L_5 \beta_5} \frac{\alpha_5 X}{\alpha_5 S_5} \times 100 \approx 20\%$$

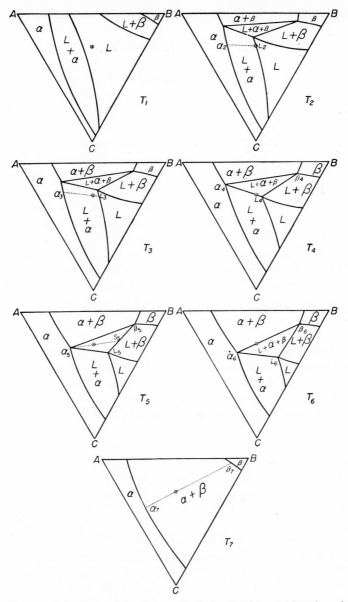

Fig. 13-8. Illustrating the sequence of equilibria involved in the freezing of an alloy whose gross composition is indicated by the circled cross in each isotherm.

133

At $T_6$ the last of the liquid disappears and there remains only

$$\%\alpha_6 = \frac{X\beta_6}{\alpha_6\beta_6} \times 100 \approx 75\%$$

and $$\%\beta_6 = \frac{\alpha_6 X}{\alpha_6\beta_6} \times 100 \approx 25\%$$

At $T_7$ the compositions of the two solid phases have changed slightly in response to the curvature of the $\alpha$ and $\beta$ solvus surfaces, and their relative proportions will again be given by the tie-line $\alpha_7\beta_7$.

Freezing has proceeded in two steps, neither of which is isothermal; primary freezing of the $\alpha$ phase has been followed by a secondary separation of $\alpha + \beta$ over the temperature range $T_4$ to $T_6$. This process is illustrated schematically in Fig. 13-9, which represents the freezing of a lead-antimony-bismuth alloy (90% lead + 5% antimony + 5% bismuth), the constitution of which is given by the space diagram in Fig. 13-10 and the isotherms in Fig. 13-11. The first five pictures of Fig. 13-9 are sketches showing the presumed progress of freezing at the temperatures indicated and corresponding to the requirements of the isotherms at the same temperatures, Fig. 13-11, where the gross composition of the alloy is indicated by the point $X$ in each section. The sixth picture ("20°C") is an actual photomicrograph of this alloy. Large white areas represent the liquid phase, unbroken black areas the primary $\alpha$, and black mixed with white designates the $\alpha$ (black) and $\beta$ (white) of the secondary constituent. At 300°C the primary crystallization of $\alpha$ (black specks) has just begun, and at 235°C these particles have achieved their maximum growth. Thereafter $\alpha$ and $\beta$ crystallize together, so that at 230°C more than half of the remaining liquid has been converted to the secondary two-phased solid constituent. Freezing is complete at 225°C, the only subsequent change being a slight increase in the volume of the $\beta$, at the expense of $\alpha$, as the compositions of the solid phases change along the solvus. The final microstructure, at 20°C, is not visibly different from that of a binary hypoeutectic alloy, but the manner of its formation differs in that the secondary constituent (the "eutectic constituent") has grown over a range of temperature instead of isothermally.

Wherever transformation occurs over a temperature range at a finite rate (i.e., natural freezing), it is anticipated that the products of that transformation will exhibit the *coring effect*. Thus, both the primary and secondary constituents, in the above example, should be cored. Coring of the primary constituent has been described in the preceding chapter; the primary $\alpha$ phase of the present example will freeze in the same manner as the $\alpha$ phase of an isomorphous system. When $\alpha$ and $\beta$ are crystallizing together, both phases should be cored (see Fig. 13-12). On this diagram, "equilibrium" freezing would begin at $T_1$, where $x_1$ falls on the line

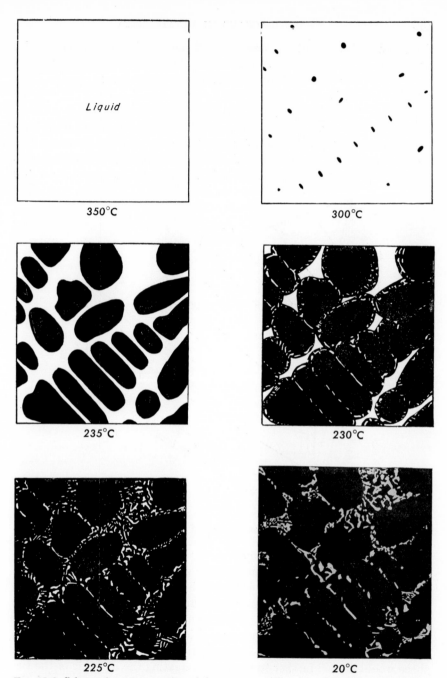

Liquid

350°C

300°C

235°C

230°C

225°C

20°C

FIG. 13-9. Schematic representation of the progress of freezing of an alloy composed of 90% Pb + 5% Bi + 5% Sb, indicated by point $x$ in Fig. 13-11. The first five pictures of the series are idealized sketches; the sixth picture is an actual photomicrograph of the alloy at room temperature. Magnification 100.

$L_1\alpha_1$, and would end at $T_3$, where $x_3$ falls on the line $\alpha_3\beta_3$. The $\alpha$ and $\beta$ phases fail to maintain equilibrium composition in natural freezing,[1] however, and their average compositions follow the lines: $\alpha_1\alpha_2'\alpha_3'\alpha_4'$ and $\beta_1\beta_2'\beta_3'\beta_4'$, respectively. Therefore, at $T_3$ some liquid remains, because $x_3$ lies within the triangle $L_3\alpha_3'\beta_3'$. At $T_4$ the gross composition point $x_4$ finally lies upon the line $\alpha_4'\beta_4'$, and crystallization is completed. This temperature is well below the final temperature of equilibrium freezing, and the composition of the last layer of each solid phase ($\alpha_4$ and $\beta_4$) lies far to the right of the gross composition point. Coring of the secondary $\alpha + \beta$

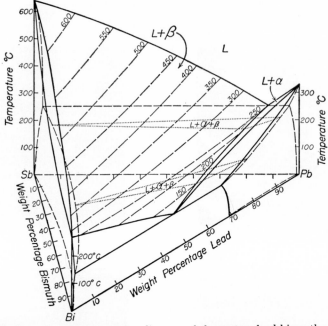

Fig. 13-10. Temperature-composition diagram of the system lead-bismuth-antimony.

constituent, although necessarily present, is rarely observed in the microstructure, partly because of the fineness of the particles into which the two phases are divided. When coring of the primary constituent precedes coring of the secondary constituent, the average composition of the solid phase concerned, in both the primary and secondary constituents, will tend toward the initial primary composition. For example, in Fig. 13-12, if the primary constituent is $\alpha$, the average composition path of the $\alpha$ phase would lie still farther to the left, elongating the $\alpha'\beta'$ side of the tie-triangle. Alloys of gross composition wholly outside the range of the

[1] For simplicity it is assumed again that the equilibrium composition of the liquid phase is maintained.

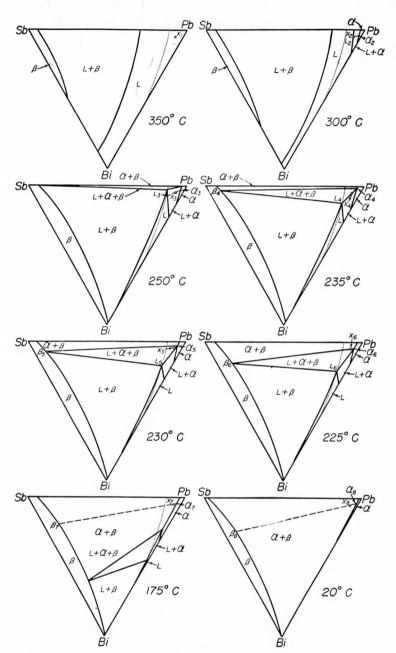

Fig. 13-11. Isotherms of the system lead-bismuth-antimony (see Fig. 13-10).

three-phase field can, therefore, undergo three-phase reaction, just as binary hypoeutectic alloys outside the eutectic range can undergo eutectic reaction.

## Heat Treatment

Structural changes accompanying heat treatment may likewise be followed by reference to the isothermal sections. Consider, for example, an alloy which at an elevated temperature, say $T_4$ in Fig. 13-6, should lie in

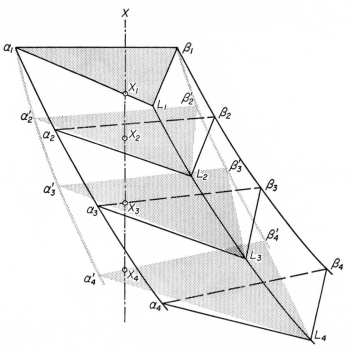

FIG. 13-12. Illustrating the origin of coring in two solid phases ($\alpha$ and $\beta$) that freeze simultaneously.

the $\alpha$ field but which, owing to nonequilibrium freezing, contains some of the $\alpha + \beta$ constituent in its structure. This alloy can be converted wholly to $\alpha$ by a homogenizing heat treatment, such as was described in Chap. 4. If at lower temperature this alloy passes into the $\alpha + \beta$ field, then a Widmanstätten precipitate of the $\beta$ phase will appear during slow cooling to room temperature or upon reheating to an intermediate temperature subsequent to quenching from the homogenizing heat-treating temperature. In almost all respects the process is analogous to that described for the binary case, differing only in that the compositions of the phases concerned lie within the ternary space diagram and must be analyzed by the use of tie-lines which appear in a sequence of isothermal sections.

## Another System Involving Three-phase Equilibrium

If two binary peritectic systems are substituted for the two binary eutectic systems of the foregoing example, the resulting ternary space model will appear as in Fig. 13-13. An exploded model of this case is shown in Fig. 13-14. This diagram has the same number of fields as was found in the previous example, and they are similarly designated. The most evident difference is to be found in the $L + \alpha + \beta$ field, which has been inverted to produce peritectic reaction; the trace of the $\alpha$ compositions on this field, line 2 in Fig. 13-14, is on the underside of the three-phase region and, therefore, is dotted in the sketch.

As in the previous example, the one-phase regions touch the three-phase region only along edges, the liquid field along edge 1, the $\beta$ field along edge 3, and the $\alpha$ field along edge 2. Two-phase regions terminate upon the three surfaces of the three-phase space. The bottom face of the $L + \beta$ region rests upon the top face 1-3-4-5 of the $L + \alpha + \beta$ field. This field is otherwise bounded by liquidus and solidus surfaces and by the confining walls of the diagram. An upper face of the $L + \alpha$ region 1-2-4-5 coincides with one of the two lower surfaces of the $L + \alpha + \beta$ region, and this field is likewise further bounded by the liquidus, solidus, and walls of the diagram. The third surface of the three-phase region rests upon the top face of the $\alpha + \beta$ region 2-3-4-5, which field is otherwise bounded by two solvus surfaces and the walls of the diagram.

These relationships are clearly evident in the isotherms of Fig. 13-15, which should be compared with those of the previous example, Fig. 13-6; it will be seen that the tie-triangles are reversed. The course of freezing may again be followed by the use of the isothermal sections. That peritectic reaction results from the reversal of the tie-triangles can be seen by analyzing the process in detail. Consider first an alloy of composition $X$ shown on tie-triangles at three temperatures in Fig. 13-16. At $T_2$ the primary separation of $\beta$ crystals is complete and the precipitation of $\alpha$ is about to begin. The quantity of primary $\beta$ is

$$\%\beta_2 \text{ (primary)} = \frac{L_2X}{L_2\beta_2} \times 100 \approx 30\%$$

When the temperature falls to $T_3$, the quantity of $\beta$ is reduced to

$$\%\beta_3 \text{ (total)} = \frac{mX}{m\beta_3} \times 100 \approx 20\%$$

and this change is accompanied by a corresponding decrease in the quantity of liquid:

$$\%L_2 = \frac{X\beta_2}{L_2\beta_2} \times 100 \approx 70\%$$

$$\%L_3 = \frac{m\alpha_3}{L_3\alpha_3}\frac{X\beta_3}{m\beta_3} \times 100 \approx 60\%$$

and an increase in the amount of the $\alpha$ phase:

$$\%_0\alpha_2 = 0$$

$$\%_0\alpha_3 = \frac{mL_3}{\alpha_3L_3}\frac{X\beta_3}{m\beta_3} \times 100 \approx 20\%$$

Just as in binary peritectic reaction, the primary constituent is consumed by reaction with the liquid to form the secondary constituent. At $T_4$ the $\beta$ phase has been consumed altogether and only liquid and $\alpha$ remain.

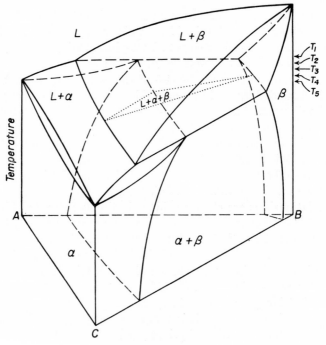

Fig. 13-13

Had the alloy been of composition $Y$, Fig. 13-16, the same course of events would have been observed, except that all the liquid would have been consumed in forming $\alpha$, leaving a residue of unreacted $\beta$. In natural freezing these processes are retarded by failure to maintain equilibrium. All constituents are cored, envelopment occurs, and the relative quantities of the crystalline phases deviate from the "ideal," just as in binary

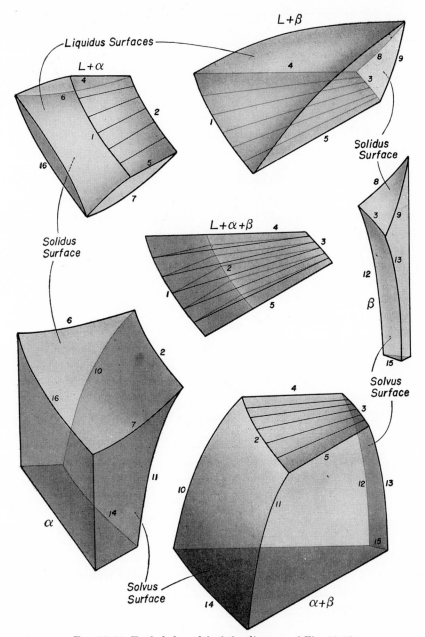

Liquidus Surfaces

$L+\alpha$

$L+\beta$

Solidus Surface

Solidus Surface

$L+\alpha+\beta$

$\beta$

Solvus Surface

$\alpha$

Solvus Surface

$\alpha+\beta$

Fig. 13-14. Exploded model of the diagram of Fig. 13-13.

141

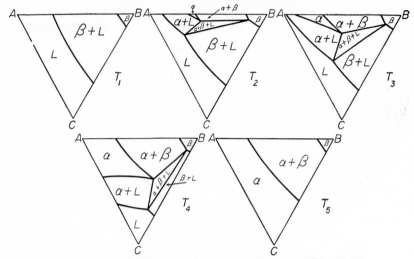

FIG. 13-15. Isotherms through the space diagram of Fig. 13-13.

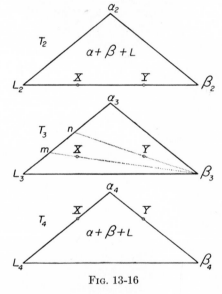

FIG. 13-16

peritectic alloys. The microstructures of cast ternary alloys of this kind are, in fact, indistinguishable from their binary counterparts, the binary peritectic alloys.

## Other Kinds of Three-phase Equilibrium

The forms that have been discussed above apply to all cases of three-phase equilibrium. Ternary three-phase eutectoid reaction is represented by the same construction as is eutectic reaction, simply substituting a solid phase for the liquid. Correspondingly, the monotectic reaction resembles the eutectic case, while the peritectoid and syntectic reactions resemble the peritectic case. *Every isothermal three-phase reaction occurring in any of the binary systems is associated with a corresponding nonisothermal three-phase reaction, described by tie-triangles, in the ternary diagram.*

With any of the above reactions the temperature of ternary three-phase equilibrium may either rise or fall as the gross alloy composition recedes

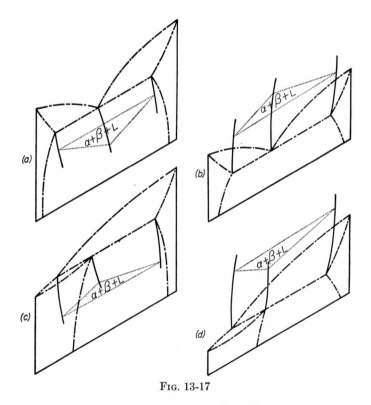

FIG. 13-17

from the binary face of the diagram (Fig. 13-17). If the tie-triangle is considered as an arrow pointing in the direction of the junction of its shorter sides, it may be said that the tie-triangle in eutectic-type reaction points away from the binary face of the ternary diagram if three-phase reaction proceeds to lower temperature, Fig. 13-17*a*, and points toward the binary face if three-phase reaction goes to higher temperature, Fig. 13-17*b*. Conversely, in the peritectic-type reactions, the tie-triangle points toward the binary side if the temperature of reaction decreases, Fig. 13-17*c*, and away if the temperature of reaction increases, Fig. 13-17*d*.

## Some Details Concerning Three-phase Equilibrium

Any pair of binary three-phase equilibria that involve the same three phases may be joined in the ternary space model by a three-phase field composed of tie-triangles. This means that a binary eutectic may be joined with a binary peritectic, a eutectoid with a peritectoid, and a monotectic with a syntectic. Where a eutectic-type system is thus joined with a peritectic-type system (Fig. 13-18), the direction of the tie-triangle

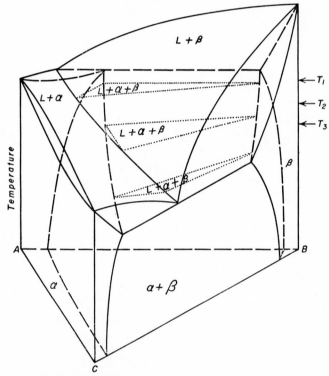

FIG. 13-18. Illustrating the rotation of the tie-triangle in passing between a binary eutectic and a binary peritectic.

changes in crossing the ternary space model. This is shown by the dotted triangles in Fig. 13-18 and also by the isotherms of Fig. 13-20. Alloys low in the $A$ component will freeze by eutectic reaction, those low in the $C$ component by peritectic reaction, and those of intermediate composition will follow a compromise course of freezing. In the present example, the intermediate alloys would at first exhibit some peritectic re-solution of the primary $\beta$ constituent; this process would taper off to a reversal where the $\alpha$ and $\beta$ phases are codeposited until the liquid is consumed. Since both phases of the secondary constituent in ternary alloys exhibit coring,

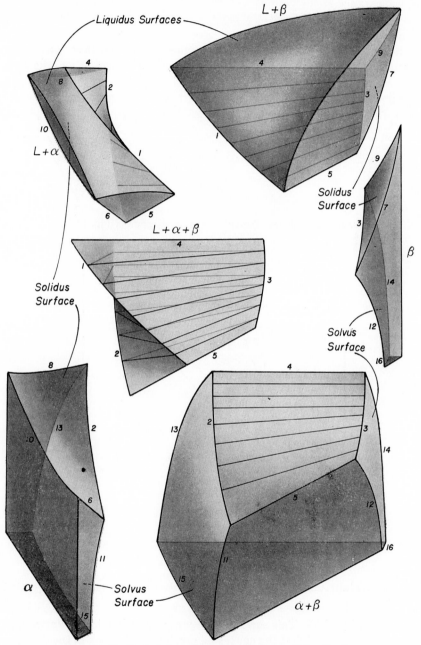

Fig. 13-19. Exploded model of the diagram of Fig. 13-18.

145

the appearance of the microstructure is altered gradually from the essentially eutectic- to the essentially peritectic-type alloys without any abrupt change of any kind.

The three-phase region of the space model has one face that turns through 180° in passing from one binary system to the other, the face

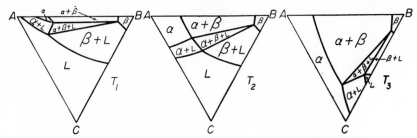

FIG. 13-20. Isotherms through the space diagram of Fig. 13-18.

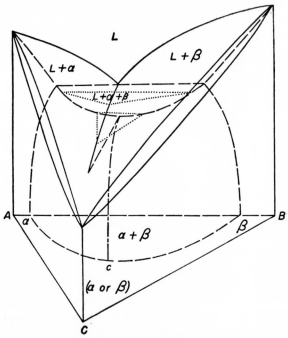

FIG. 13-21. Illustrating the termination of a three-phase equilibrium in a critical tie-line.

1-2-4-5 in Fig. 13-19. This surface is only partly visible in the drawing of the three-phase field, but the concealed portion can be seen in the drawing of the $L + \alpha$ field. In other respects this diagram closely resembles the two cases previously discussed.

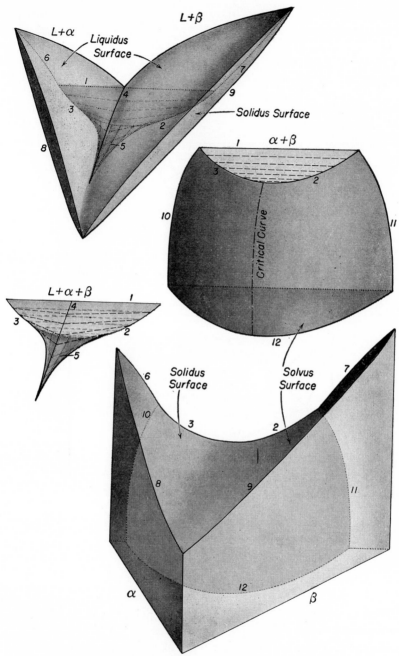

FIG. 13-22. Exploded model of the diagram of Fig. 13-21.

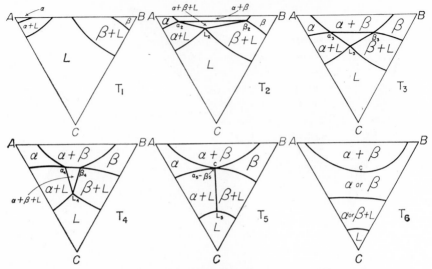

Fig. 13-23. Isotherms through the space diagram of Fig. 13-21.

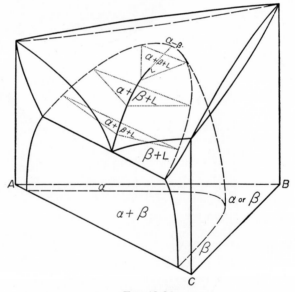

Fig. 13-24

In all the examples considered up to this point the three-phase region has terminated upon three-phase isotherms in the binary faces of the space model. There are two other ways in which the three-phase region may be terminated. One of these involves junction with a four-phase reaction, a subject which will be discussed in succeeding chapters; the other is termination upon a critical tie-line in the ternary model (see Fig. 13-21). Here the $L + \alpha + \beta$ region begins upon the binary eutectic line, and the compositions of $\alpha$ and $\beta$ approach each other as the temperature falls (Fig. 13-23) until, at a critical point $c$ $(T_5)$, the two solid phases become indistinguishable. In this way, the tie-triangle closes to a tie-line $\alpha_5$-$\beta_5 L_5$ and the three-phase region ends.

The three-phase field of this example has two sharply curved surfaces, 1-3-4-5 and 1-2-4-5 in Fig. 13-22, and one surface of lesser curvature, 1-2-3. These join to produce a shape which is reminiscent of a bird's beak. The pointed part of the "beak" extends beyond the $\alpha + \beta$ region and penetrates the solid plus liquid field.

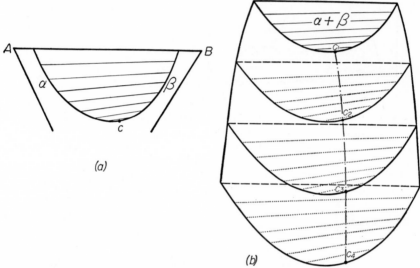

FIG. 13-25. A critical point occurs at $c$ in each isotherm; the trace of the critical points in a series of isotherms produces the critical curve $c_1 c_4$.

When the temperature of three-phase equilibrium rises as the composition recedes from the binary eutectic, the $L + \alpha + \beta$ field is inverted (Fig. 13-24) and ceases to resemble a beak; it terminates upon the tie-line $L\alpha$-$\beta$.

## Critical Points and Critical Curves

It will be apparent that the merging of the $\alpha$ and $\beta$ phases in the foregoing diagram corresponds to a critical point in two-phase equilibrium. At every temperature from $T_5$ downward in Fig. 13-23, the isotherm of the $\alpha + \beta$ field exhibits a critical point $c$, as is shown in detail in Fig. 13-25$a$. The tie-lines of the $\alpha + \beta$ field become progressively shorter as the composition moves away from the $AB$ side of the diagram, reaching zero length at point $c$, the *critical point*. Since there is a critical point at each temperature level in the two-phase field, these points trace a *critical curve* $c_1c_2c_3c_4$ in Fig. 13-25$b$, also shown as a dash-dot line on the boundary

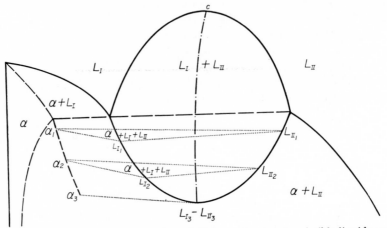

Fig. 13-26. Termination of a three-phase field involving two immiscible liquids and a solid phase at a critical point and tie-line.

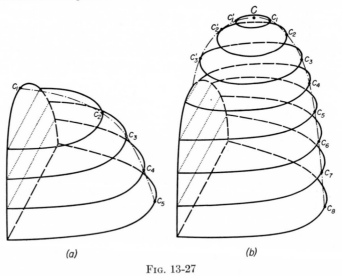

(a)　　　　　　　　(b)

Fig. 13-27

of the $\alpha + \beta$ field in Figs. 13-21 and 13-22. The $\alpha$ and $\beta$ phases are distinguishable only when they occur together in the $\alpha + \beta$ region or the $L + \alpha + \beta$ region. Elsewhere they form continuous series of solid solutions as in an isomorphous system. Obviously, the "two" solid phases must have the same crystal structure.

Another common example of the ending of a three-phase field upon a tie-line within the space model involves monotectic reaction (Fig. 13-26). Here the field $L_I + L_{II} + \alpha$ begins at the binary monotectic line and ends upon the tie line $\alpha_3 L_{I_3}$-$L_{II_3}$. Again the tie-triangle has closed to a line at a critical point $L_{I_3}$-$L_{II_3}$, where the two liquids become indistinguishable. The $L_I + L_{II}$ region in the space model has a critical curve $cL_{I_3}$-$L_{II_3}$.

It does not always happen that the highest point (temperature maximum) of the critical curve lies in a binary face of the space model; sometimes there is a ternary critical point which lies above the binary critical point. The two cases are illustrated in Fig. 13-27, where the "highest" critical point occurs in the binary face in drawing $a$ and at point $C$ within

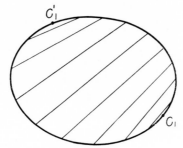

FIG. 13-28. Isotherm above the binary critical point in Fig. 13-27$b$.

the ternary space of the diagram in drawing $b$. In the latter case there are two critical points in each isotherm at any temperature lying between the ternary and binary critical points (Fig. 13-28).

## Maxima and Minima

The three-phase region may pass through a maximum or minimum in temperature. When it does so, *the tie-triangle becomes a tie-line* connecting the three phases that are in equilibrium at the temperature maximum or minimum (Figs. 13-29 and 13-31). The details of these diagrams may be understood more easily by reference to the corresponding isotherms in Figs. 13-30 and 13-32. These examples involve only eutectic reaction; it is equally possible, however, to have maxima and minima with peritectic-type reactions or with combinations of eutectic- with peritectic-type reactions, and so on. Attention is directed to the analogy between the tie-triangle, which closes to a single line at a temperature maximum or minimum, and the tie-line of two-phase equilibrium which is reduced to a point at a temperature maximum or minimum, as in Fig. 12-18. In both cases the transformation assumes the form of a binary univariant reaction.

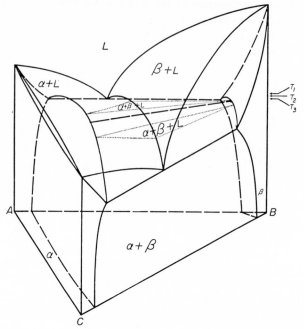

FIG. 13-29. Three-phase region reduces to a tie-line at a temperature maximum.

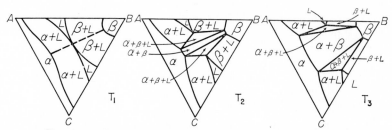

FIG. 13-30. Isotherms through the space diagram of Fig. 13-29.

## Vertical Sections

Although bearing some resemblance to binary diagrams, vertical sections through the ternary space models discussed in this chapter have several very important differences that distinguish them from binaries. The reaction horizontal of the binary diagram is replaced by a three-phase region which has no fixed form, no straight or horizontal boundaries except by coincidence, and no definite number of bounding curves. Two simple cases are presented in Figs. 13-33 and 13-34. These sections correctly record the transformation temperatures for alloys having gross

compositions on the lines $XB$ and $WB$, respectively; they do not, however, show the compositions of the phases involved in the corresponding
equilibria, except in the one-phase fields. For example, an alloy of intermediate composition, held at a temperature within the $\alpha + \beta$ field, will

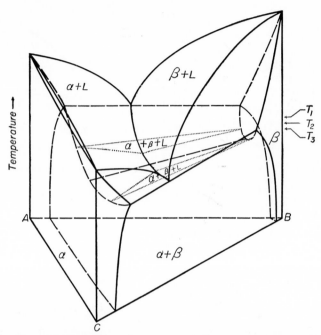

Fig. 13-31. Three-phase region reduces to a tie-line at a temperature minimum.

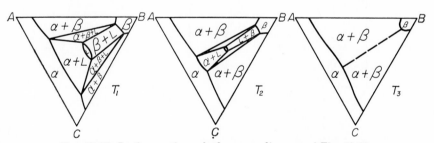

Fig. 13-32. Isotherms through the space diagram of Fig. 13-31.

be composed of $\alpha + \beta$, but the compositions of the conjugate phases cannot be read from the vertical section, nor can the lever principle be
applied to ascertain the relative quantities of the two phases present at
equilibrium. This situation is even more apparent in section $YZ$ of Fig.
13-35, where the $\beta$ phase is named in three different fields and yet does not
appear, as such, at any point on the section. All compositions of $\beta$ lie out-

side the composition range of this section. Another striking example of this is shown in Fig. 13-36. The section $RS$ resembles section $XB$ of Fig. 13-33, but in section $TU$ the $L + \alpha + \beta$ field is entirely detached from the solid field and is bounded by a continuous line.

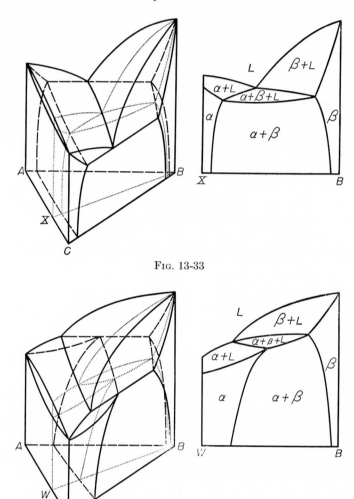

Fig. 13-33

Fig. 13-34

Despite these disadvantages, vertical sections are usefully employed for the representation of portions of ternary alloy systems lying close to one of the binary sides of the space diagram. Many examples exist among iron-base ternary alloy systems. Of particular interest is the influence of a third element upon the iron-carbon eutectoid reaction. With the first addition of a third element this reaction, of course, ceases to be isothermal, even under equilibrium conditions, but occurs over a temperature range.

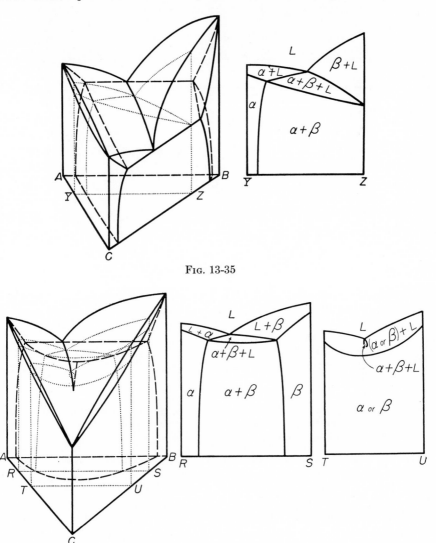

Fig. 13-35

Fig. 13-36

The temperature range of transformation is elevated by some elements, lowered by others. These modifications change the rate of transformation, requiring a separate TTT curve for each individual alloy of each ternary alloy system.

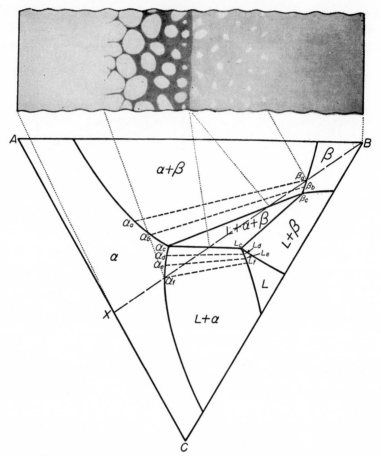

FIG. 13-37. Schematic representation of the relation between the structure of a ternary diffusion couple and the phase diagram. Layers are developed corresponding to both one- and two-phase regions lying upon the composition path between the extremes of composition; three-phase regions correspond to layer interfaces in the diffusion couple. The composition path of the diffusion couple is not usually straight as it is shown to be in this ideal example.

## Isothermal Diffusion in Ternary Systems

Practical examples of approximately isothermal diffusion among three or more metals are far more numerous than those involving only two metals. Such processes as soldering, brazing, galvanizing, calorizing, oxidizing alloys at high temperatures, sintering complex metal powder mixtures, and many others involve substantially isothermal diffusion simultaneously among more than two metals. In view of its potential value, therefore, it is unfortunate that information with regard to the relationships existing between ternary constitution and the structures produced by diffusion among three elements is very scanty. At the present time it is possible to say only that ternary diffusion can result in the formation of both one-phase and two-phase layers in a diffusion couple, but not three-phase layers. As with binary diffusion, no layers at all, but simply interfaces between layers, correspond to regions of divariant (or uni- or invariant) equilibrium.

These relationships are illustrated schematically in Fig. 13-37. A block of an alloy of composition $X$ has been held in contact with a block of metal $B$, at the temperature of the isotherm, until diffusion layers have developed, as shown by the sketch above. It is assumed that all gross compositions that would be found by analyzing vertical slices through the diffusion sample, taken perpendicular to the plane of the drawing, would lie upon the line $XB$. This is known to be an erroneous assumption, but it provides the best approximation that can be made at present. The line $XB$ crosses the fields $\alpha$, $L + \alpha$, $L + \alpha + \beta$, $\alpha + \beta$, and $\beta$. Presumably, the $\alpha$ layer varies in composition from $X$ to $\alpha_f$ and the $\beta$ layer from $\beta_a$ to $B$. The $\alpha$ of the $L + \alpha$ layer should then range in composition from $\alpha_f$ to $\alpha_c$, and the conjugate liquid from $L_f$ to $L_c$, while in the $\alpha + \beta$ layer the $\alpha$ should vary from $\alpha_c$ to $\alpha_a$ and the $\beta$ from $\beta_c$ to $\beta_a$. The $L + \alpha + \beta$ region is represented in the diffusion sample by the interface between the zones of $L + \alpha$ and $\alpha + \beta$. This course of reasoning is more or less successful in predicting the qualitative nature of "ternary diffusion structures," but it does not ordinarily predict the correct compositions of the several phases. In some cases where the diffusion velocities of the three components are markedly different, the path of composition change deviates radically from the direct path between the compositions of the end points.

## Physical Properties of Alloys

Structurally, the ternary alloys of systems such as have been discussed in this chapter differ from the corresponding binary alloys chiefly in that "coring" is to be expected in cast ternary alloys in both the primary and secondary constituents. This may be expected to have only a minor influence upon the physical properties. The individual phases of the structure become ternary, rather than binary, solid solutions, and their particle sizes may be altered. These factors are capable of exerting a pronounced influence upon the physical properties, but no general principles governing the direction or magnitude of such effects are known at present. It might be supposed that adding a third element to each solid phase would be the equivalent of increasing the amount of material in solid solution and would, generally, increase the hardness and strength of the phase. This does not always happen, however, because the boundary surfaces of the one-phase fields sometimes curve in such manner that the total quantity of substance in solid solution in a given phase is actually less in the ternary than in the binary alloys.

## PRACTICE PROBLEMS

**1.** The binary systems $AB$ and $BC$ are both like that of Fig. 5-1, being isomorphous through the freezing range and of the eutectoid type at lower temperature. The binary system $AC$ is isomorphous at all temperatures, i.e., both above and beneath the temperatures of allotropic transformation of $A$ and $C$. Draw the space (TXY) diagram of the ternary system $ABC$; develop isotherms sufficient in number to display the main features of the internal structure of this diagram; develop isopleths through the $B$ corner and also parallel to the $AC$ side of the diagram.

**2.** Consider an alloy of gross composition near the mid-point of the eutectic valley in Fig. 13-21. Deduce the course of natural freezing of this alloy. In what ways might the cast alloy be expected to respond to heat treatment?

**3.** The binary system $AC$ is isomorphous, the binary system $BC$ is of the monotectic type, and the binary system $AB$ is of the syntectic type. Draw a space diagram of the ternary system $ABC$, and check its validity by developing isotherms. The simplest solution of this problem results when it is assumed that the solid is isomorphous beneath the monotectic in $BC$ and beneath the syntectic in $AB$. After solving the problem in this simpler form, substitute the binaries of Figs. 3-2, 6-1, and 10-1 and repeat.

# CHAPTER 14

# TERNARY FOUR-PHASE EQUILIBRIUM—CLASS I

Ternary eutectic reaction occurs by the "isothermal" decomposition of liquid into three different solid phases:

$$L \underset{\text{heating}}{\overset{\text{cooling}}{\rightleftharpoons}} \alpha + \beta + \gamma$$

This is known also as *four-phase equilibrium of the first kind* or as an example of *class I four-phase equilibrium*. It is represented in the space model by a unique tie-triangle, the ternary eutectic isotherm (or plane), which connects the compositions of the four phases participating in the reaction at the ternary eutectic temperature (Fig. 14-1). In this diagram the ternary eutectic plane has been shaded to make its location more apparent. The three corners of the triangle touch the three one-phase regions $\alpha$, $\beta$, and $\gamma$ and are so labeled. The liquid composition occurs at point $L$ within the triangle, where the liquidus surfaces from the three corners of the space diagram meet at the lowest melting point of the ternary system.

According to the phase rule, four-phase equilibrium in ternary systems should be univariant:

$$P + F = C + 2$$
$$4 + 1 = 3 + 2$$

Having established the pressure, the temperature of four-phase equilibrium and the compositions of each of the four phases should be fixed. The construction employed in Fig. 14-1 meets these requirements; the ternary eutectic plane is isothermal, and the compositions of the four phases are designated at four fixed points on the eutectic plane.

A fuller understanding of the internal structure of the ternary eutectic diagram may be had by reference to the "exploded model" in Fig. 14-2. The diagram is qualitatively symmetrical with respect to each of its three corners or three sides, there being only six different kinds of fields, of which all except the ternary eutectic plane have been encountered in previous chapters.

Three three-phase fields $L + \alpha + \beta$, $L + \alpha + \gamma$, and $L + \beta + \gamma$, Fig. 14-2, emerge downward from the three corresponding binary eutectic reactions, Fig. 14-1, and terminate upon the ternary eutectic plane $L + \alpha + \beta + \gamma$. The bottom tie-triangle in each of these fields joins those in the other two three-phase fields to form one larger tie-triangle, the ternary eutectic plane. Thus triangle 1-5-3 of the $L + \alpha + \beta$ field, triangle 1-2-4 of the $L + \alpha + \gamma$ field, and triangle 2-3-6 of the $L + \beta + \gamma$ field join to form triangle 4-5-6 of the ternary eutectic plane. Below the ternary eutectic temperature the three-phase field $\alpha + \beta + \gamma$ begins with the tie-triangle 4-5-6 and descends in a series of tie-triangles to the base of the space diagram.

Surmounting the three-phase fields are three two-phase fields $L + \alpha$, $L + \beta$, and $L + \gamma$. These are bounded by liquidus and solidus surfaces connected at all points by tie-lines joining the conjugate liquid and solid compositions. Each is wedge-shaped on its underside, having two ruled surfaces which are, respectively, identical with one top surface on each of two three-phase fields. For example, the region $L + \beta$ has an undersurface 12-9-3-17 composed of tie-lines joining the conjugate liquid and $\beta$ phases along the lines 9 and 17, respectively, and being identical with the surface 12-9-3-17 on the $L + \beta + \gamma$ field. Its other lower surface is 3-8-11-16, which coincides with the surface of like designation on the $L + \alpha + \beta$ field. Line 3, it will be noted, lies in the ternary eutectic plane and connects the liquid composition at the eutectic point with the $\beta$ of the ternary eutectic. Similar relationships will be observed upon the $L + \alpha$ and $L + \gamma$ fields.

There are also three two-phase fields representing the coexistence of three pairs of solid phases: $\alpha + \beta$, $\alpha + \gamma$, and $\beta + \gamma$. These are slablike fields bounded on their vertical ends by solvus surfaces between which run tie-lines connecting pairs of conjugate solid phases. Upon the $\alpha + \beta$ field, Fig. 14-2, for example, the solvus surfaces are 26-15-23 and 24-16-31. The upper surface of this field 5-15-11-16 is a ruled surface composed of tie-lines joining compositions of $\alpha$ along line 15 with compositions of $\beta$ along line 16; this surface is identical with the ruled undersurface of the $L + \alpha + \beta$ field 5-15-11-16. Another ruled surface 34-23-5-24 forms the front face of the $\alpha + \beta$ region, again being composed of tie-lines connecting $\alpha$ and $\beta$ compositions along the lines 23 and 24 and coincident with the rear side of the $\alpha + \beta + \gamma$ field. These constructions are duplicated in the $\alpha + \gamma$ and $\beta + \gamma$ fields.

One-phase regions fill the remaining space of the ternary diagram. Above the three segments of the liquidus surface and extending down to the ternary eutectic point lies the region of liquid. At each corner of the space diagram is a one-phase field representing one of the three crystalline phases. These are represented in Fig. 14-2 and require no special comment except to point out that there is one corner on each (labeled $\alpha$, $\beta$, and $\gamma$ on the respective fields) which touches the ternary eutectic plane and there are upon each one-phase field three edges (lines 14, 15, and 23 on the $\alpha$ field) which coincide with edges on three of the three-phase fields.

Viewed through the medium of isothermal sections, the essential simplicity of the ternary eutectic diagram is even more apparent. The seven isotherms of Fig. 14-3 are taken at the temperatures designated $T_1$ to $T_7$ in Fig. 14-1. In the first section $T_1$ the liquidus and solidus adjacent

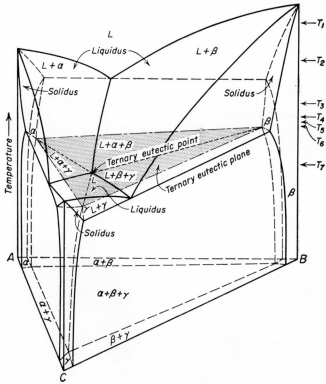

Fig. 14-1. Temperature-composition space diagram of a ternary eutectic system, illustrating class I four-phase equilibrium.

to $B$, the highest melting component, are intersected, revealing fields of $L$, $\beta$, and $L + \beta$. Successively at $T_2$ and $T_3$ the liquidus and solidus pairs adjacent to the $A$ and $C$ components are intersected. The third section $T_3$ is taken just below the eutectic temperature of the binary system $AB$ and intersects the three-phase region $L + \alpha + \beta$ that originates upon the binary eutectic line. Similar constructions were found in each of the phase diagrams of the preceding chapter; for example, compare with section $T_2$ in Fig. 13-6. From each of the binary eutectic lines there issues a three-phase field so that at $T_4$, just below the binary eutectic of the system $AC$, the field $L + \alpha + \gamma$ appears and at $T_5$, just below the $BC$ eutectic, the $L + \beta + \gamma$ field appears. The liquid region is now confined to a small three-cornered area in the middle of the diagram, and the liquid plus solid regions have grown narrow. As the temperature falls, these fields continue to shrink, permitting the three three-phase tie-triangles to meet and to form the ternary eutectic plane at $T_6$, the ternary eutectic temperature. At this temperature the liquid phase disappears and with it the six other fields involving liquid, so that below the ternary eutectic temperature, only solid phases remain. As the extent of solid solubility decreases with falling temperature $T_7$, the $\alpha + \beta + \gamma$ field grows larger while the one-phase fields contract.

### Vertical Sections

Vertical sections through the ternary eutectic diagram produce a wide variety of forms. The ternary eutectic plane appears only as a horizontal line in such sections; otherwise there is little regularity. Section $XB$ of Fig. 14-4 has been so selected that it intersects the $B$ corner of the space diagram and the ternary eutectic point. There are no tie-lines in this section, and all boundaries are curved, except the trace of the ternary eutectic plane. The next section, $YZ$, Fig. 14-5, has been caused to include

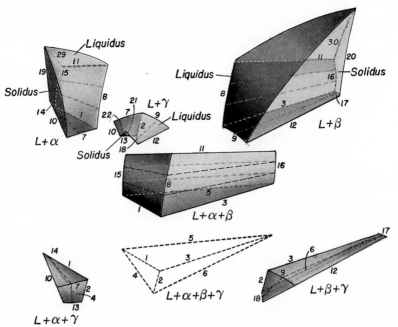

Fig. 14-2. Exploded model of the diagram of Fig. 14-1. The inscribed numbers

that line in the ternary eutectic plane which joins the composition of the liquid phase with that of the $\beta$, that is, the lowest tie-line of the $L + \beta$ field. Here the $L + \alpha + \beta$ field is not intersected and the $L + \beta$ field ends on the ternary eutectic line. Two sections parallel to the $AB$ side of the space model are shown in Figs. 14-6 and 14-7; neither passes through the ternary eutectic point. It will be noted that whereas the three-phase regions are most often enclosed by three boundaries in Fig. 14-6, the $L + \alpha + \gamma$ and $L + \beta + \gamma$ fields are enclosed by four boundaries each in Fig. 14-7. No tie-lines appear in either of these sections. As in previous examples, temperatures of transformation can be read from these vertical sections, but the compositions of the participating phases are shown

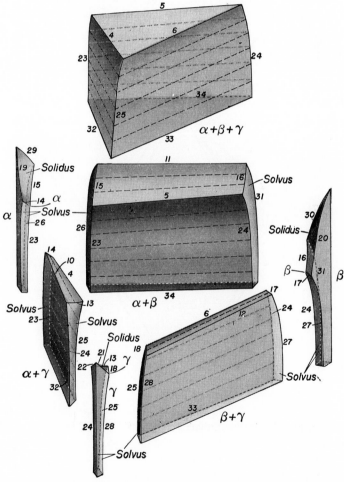

identify edges that are identical lines on different segments of the model.

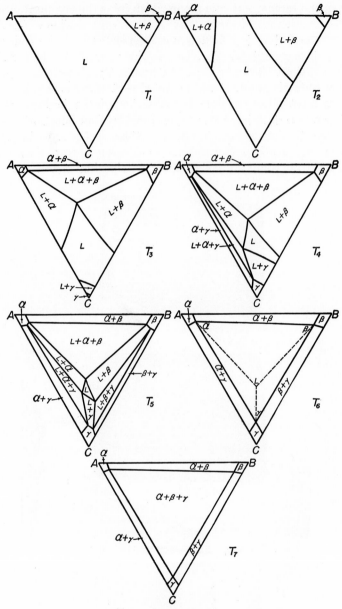

FIG. 14-3. Isotherms through the space diagram of Fig. 14-1.

164

only at a few specific points such as $L$ and $\beta$ at the ternary eutectic temperature in Fig. 14-5.

### Freezing of the Ternary Eutectic Alloy

The alloy of lowest melting point, the *ternary eutectic alloy* (point $L$ in isotherm $T_6$ of Fig. 14-3), freezes isothermally. Above the eutectic temperature, this alloy is fully molten, and below, it is entirely solid. At the

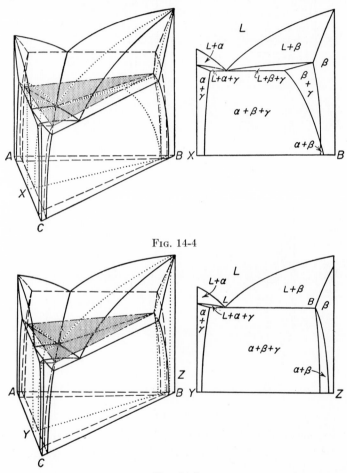

FIG. 14-4

FIG. 14-5

eutectic temperature the liquid decomposes simultaneously into crystals of $\alpha$, $\beta$, and $\gamma$. This process results in a fairly uniform distribution of the three solid phases in the microstructure of the alloy. An example of a ternary eutectic structure is given in Fig. 14-8A, which represents an alloy composed of lead plus 11% antimony and 4% tin, composition $e$ in Fig. 14-9. The lead-rich $\delta$ phase (medium gray), being present in major quantity in this eutectic, forms a matrix in which the antimony-rich $\alpha$ needles (white) and the $\beta$ needles (black) are embedded. Since $\delta$ is the

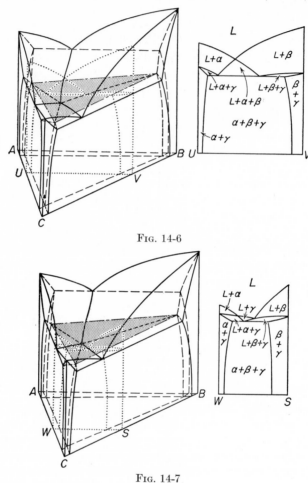

Fig. 14-6

Fig. 14-7

continuous phase, the physical properties of this alloy more closely resemble those of the lead-rich phase than those of either of the other two phases.

## Freezing of Other Alloys of the Ternary Eutectic System

When the gross composition departs from that of the eutectic, the freezing process may result in microstructures which exhibit (1) a one-phase primary constituent with a two-phase secondary constituent and a three-phase tertiary constituent, alloy $X$ in Fig. 14-9; (2) a two-phase primary constituent with a three-phase secondary constituent, alloy $Y$ in Fig. 14-9; or (3) a one-phase primary constituent with a three-phase secondary constituent, alloy $Z$ in Fig. 14-9. Photomicrographs of cast lead-tin-antimony alloys of these types are given in Fig. 14-8C, B, and D respectively.

Alloy $X$ of Fig. 14-9 begins to freeze at 250°C when primary (idiomorphic) particles of the $\beta$ phase (black) begin to separate (Fig. 14-8C). Just above 240°C this alloy enters the $L + \beta + \delta$ field and somewhat acicular particles of $\beta$ crystallize, together with the $\delta$ phase (gray), in a Chinese script form that surrounds the primary $\beta$ particles. All the liquid that remains at 240°C freezes isothermally to the ternary eutectic of $\alpha + \beta + \delta$.

This freezing process may be followed in a simplified manner by reference to a liquidus projection such as that shown in Fig. 14-10. Here the valleys of the liquidus surface are represented in solid lines, the outline of the ternary eutectic plane is dashed, and the other boundaries of the three-phase regions at the eutectic temperature are shown in dotted lines. Arrows proceeding from $X$ toward $e$ indicate the path of change of the liquid composition. At first, the liquid composition moves almost directly away from the $B$ corner; during this stage the primary $\beta$ phase is crystallizing. When the composition of the liquid reaches the valley that descends from the $AB$ eutectic to the ternary eutectic, it follows this path, $\alpha$ and $\beta$ crystallizing together meanwhile. At $e$, any remaining liquid freezes to a eutectic constituent without further change of composition.

Alloy $Y$ lies upon the valley that descends from the $\beta\delta$ eutectic to the ternary eutectic point in Fig. 14-9. Consequently, freezing begins with the coseparation of two solid phases, $\beta$ and $\delta$ (black needles in a gray matrix) in Fig. 14-8B. All liquid that remains at the ternary eutectic temperature, of course, freezes to a three-phase $\alpha + \beta + \delta$ eutectic structure. With equilibrium freezing no liquid would remain at the eutectic temperature, because composition $Y$ lies outside the eutectic triangle, but with natural freezing the coring process will result in the extension of the freezing range and some liquid will be left when the ternary eutectic temperature is reached. In Fig. 14-10 it can be seen that the liquid composition travels directly down the liquidus valley from $Y$ to $e$.

Alloy $Z$ represents a special case in which the composition lies exactly on the line connecting the $\delta$ corner of the ternary eutectic triangle with the eutectic point. This line is also the lowest tie-line in the $L + \delta$ region and the line upon which the $L + \beta + \delta$ and $L + \alpha + \delta$ regions meet. Freezing begins with a primary separation of the $\delta$ phase, Fig. 14-8D, while the liquid composition moves directly toward the ternary eutectic

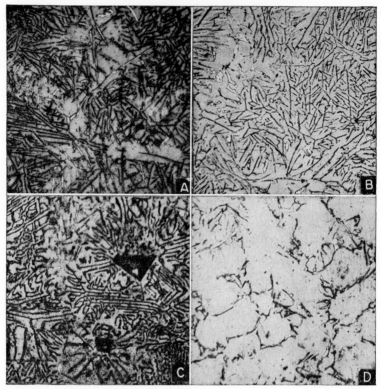

FIG. 14-8. Microstructure of four alloys of the lead-tin-antimony system in the as-cast state. The ternary eutectic alloy, shown at A (point *e* in Figs. 14-9 and 14-10), is composed of a light-colored matrix of δ supporting a large number of black β particles and a few white α particles; at B is shown an alloy the composition of which lies upon one of the liquidus valleys adjacent to the ternary eutectic (point *Y* in Figs. 14-9 and 14-10), so that primary freezing results in a two-phase crystallization of δ (light matrix) and β (black needles) followed by a secondary crystallization of the ternary eutectic. Shown at C, an alloy of random composition (point *X* in Figs. 14-9 and 14-10) freezes by the primary crystallization of large idiomorphic particles of β (black), a secondary separation of β (black needles) + α (white particles), and a tertiary freezing of the residual liquid as a ternary eutectic. Alloy D lies upon the line connecting the δ corner of the ternary eutectic plane with the eutectic liquid (point *Z* in Figs. 14-9 and 14-10) and is composed of primary δ dendrites interspersed with ternary eutectic. Magnification 50.

point *e* in Fig. 14-10. No valley of the liquidus is intersected, and no two-phase crystallization occurs. All remaining liquid is of ternary eutectic composition and freezes to the ternary eutectic constituent.

Thus, in place of the two basic types of microstructure of binary eutectic systems (the eutectic and the hypo- or hypereutectic), there are four distinct kinds of microstructure that are characteristic of ternary eutectic systems. None of these except the ternary eutectic itself is named. Under-

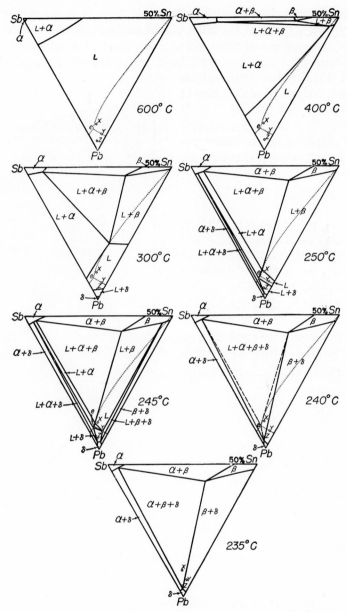

FIG 14-9. Isotherms from the ternary eutectic system Pb-Sb-β (50% Sb + 50% Sn). The letters e, X, Y, and Z indicate the compositions of the four alloys, the microstructure of which is shown in Fig. 14-8.

169

cooling is common in the freezing of ternary alloys, and as a consequence, the expected structures are often considerably modified. Also, divorcement of two-phase and three-phase constituents can lead to extensive alteration of the microstructure. This situation is further complicated by the fact that the binary three-phase reactions need not be of the eutectic types, but one or more may be peritectic type. The $L + \alpha + \beta$ equilibrium of the antimony-tin binary system is, in fact, of the peritectic type.

## Heat Treatment

No new principles are involved in the heat treatment of alloys of ternary eutectic systems. Holding at temperature sufficiently high to accelerate diffusion causes the alloys to approach their equilibrium states. The end point of any such homogenization treatment can be read from the isothermal section

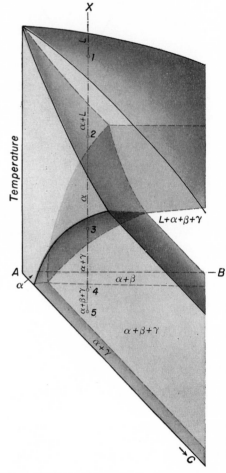

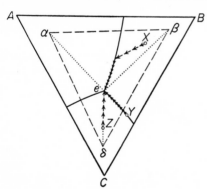

FIG. 14-10. Diagram showing by arrows the progress of composition change of the liquid phase during the freezing of alloys $X$, $Y$, and $Z$.

FIG. 14-11. Detail of the $A$ corner of a ternary eutectic space diagram showing that a typical alloy $X$ can exist in five different equilibrium states.

corresponding to the temperature of heat treatment by noting the field within which the gross composition of the alloy falls and by applying the appropriate form of the lever principle to ascertain the relative quantities of the phases that would be present at equilibrium. Since departures from the equilibrium state can be induced during freezing by coring of both

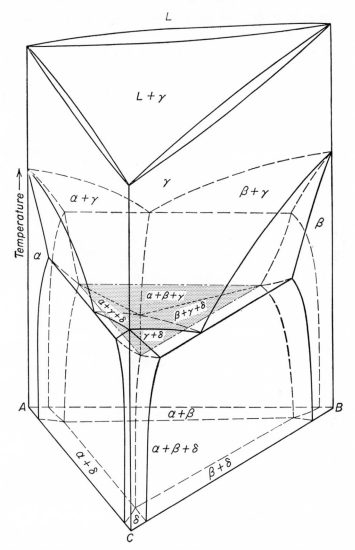

FIG. 14-12. Space diagram of an idealized ternary eutectoid system.

the primary and secondary (two-phase) constituents, and since there is also an increase in the number of changes that can be induced in the solid state by various heat-treating cycles, it is impractical to consider all possibilities here. Instead, one example involving several of the factors that are normally encountered will be discussed.

Consider an alloy of composition $X$, Fig. 14-11, which should be composed of $\alpha + \beta + \gamma$ at room temperature and up to temperature 4. Above

4 and up to 3, it would be composed of $\alpha + \gamma$, from 3 to 2 of $\alpha$ alone, from 2 to 1 of $L + \alpha$, and above 1 of liquid alone. This alloy, if normally frozen, would have a microstructure similar in kind to that of Fig. 14-8C; i.e., there would be a primary constituent of cored $\alpha$, a secondary constituent of cored $\alpha + \gamma$, and a tertiary constituent composed of the ternary eutectic $\alpha + \beta + \gamma$. If held at a temperature between 2 and 3 the

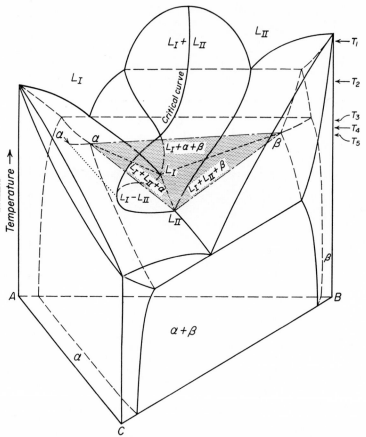

Fig. 14-13. Space diagram illustrating class I four-phase equilibrium involving two immiscible liquid phases.

$\beta$ and $\gamma$ phases would dissolve completely in the $\alpha$, coring of the $\alpha$ phase would be reduced, and the alloy, if quickly cooled from this temperature and examined, would be found to be composed of homogeneous $\alpha$ and no other phase. Subsequent reheating to a temperature between 3 and 4 would cause a Widmanstätten precipitate of $\gamma$ to form, or reheating to a temperature below 4 would cause both $\beta$ and $\gamma$ to precipitate in Widmanstätten array. This might or might not be accompanied by age-hardening effects, depending upon whether or not the conditions described in Chap. 4 are met.

If, instead of heating the cast alloy at a temperature between 2 and 3. it is held at a temperature between 3 and 4, only the $\beta$ phase will be dissolved. The $\gamma$ phase, which is present in lesser quantity, would, in all probability, become spheroidized so that if quenched, the heat-treated

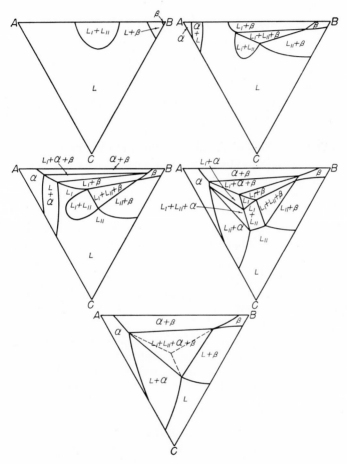

Fig. 14-14. Isotherms through the space diagram of Fig. 14-13.

alloy would be found to be made up of a matrix of homogeneous $\alpha$ surrounding rounded particles of $\gamma$. Reheating below temperature 4 would cause the appearance of a Widmanstätten precipitate of $\beta$.

Liquation of this alloy can also appear in several different ways. If the alloy is previously homogenized in the $\alpha$ field, liquation will begin upon exceeding temperature 2. If, on the other hand, the alloy were homogenized within the $\alpha + \gamma$ field, liquation would begin at some lower temperature corresponding to the bottom surface of the $L + \alpha + \gamma$ field. The as-cast alloy containing a residue of the ternary eutectic constituent will, of course, liquate at the ternary eutectic temperature unless the rate of heating is sufficiently slow to permit the $\beta$ phase to be dissolved before the eutectic temperature is reached.

### The Ternary Eutectoid and Some Other Examples of Class I Reactions

By substituting a solid phase for the liquid phase of the ternary eutectic, the *ternary eutectoid* is obtained (Fig. 14-12). The analogy between the relationship of the binary eutectic to the ternary eutectic and that of the binary eutectoid to the ternary eutectoid appears good in all respects. Relatively little is known, however, concerning the mechanism and rate of any ternary eutectoid transformation.

Many other reactions of the class I type are obtainable by combining other phases in this kind of diagrammatic construction. A list of all conceivable combinations of phases in class I reaction is given in Table 4, Chap. 18; only a few of these have actually been observed. A monotectic version of class I reaction is depicted in Figs. 14-13 and 14-14. Here the region of two-liquid immiscibility crosses a eutectic valley. At the intersection there are four phases $L_I$, $L_{II}$, $\alpha$, and $\beta$, establishing isothermal equilibrium. Three three-phase regions descend from higher temperature toward the four-phase reaction plane; these are $L_I + L_{II} + \beta$, which originates upon the binary monotectic isotherm; $L_I + \alpha + \beta$, which originates upon the $AB$ binary eutectic isotherm; and $L_I + L_{II} + \alpha$, which originates upon the critical tie-line $\alpha L_I$-$L_{II}$ (dotted in Fig. 14-13) that marks the limit of two-liquid immiscibility, as in the example of Fig. 13-26. Four-phase reaction results in the disappearance of one of the liquid phases, and below this temperature only three phases—$L_{II}$, $\alpha$, and $\beta$—exist. Below the temperature of the $BC$ binary eutectic only the solid phases $\alpha$ and $\beta$ remain.

### PRACTICE PROBLEMS

1. In a certain ternary eutectic equilibrium the compositions of the conjugate solid phases are $\alpha = 60\% \ A + 20\% \ B + 20\% \ C$, $\beta = 20\% \ A + 70\% \ B + 10\% \ C$, and $\gamma = 10\% \ A + 30\% \ B + 60\% \ C$. Ascertain by graphical means which of the following are possible compositions of the liquid phase of this ternary eutectic: (a) $50\% \ A + 20\% \ B + 30\% \ C$, (b) $40\% \ A + 40\% \ B + 20\% \ C$, (c) $20\% \ A + 60\% \ B + 20\% \ C$, (d) $10\% \ A + 40\% \ B + 50\% \ C$. What would be the percentages of $\alpha$, $\beta$, and $\gamma$ into which each "possible" liquid would separate during eutectic decomposition?

2. The ternary eutectic system $ABC$ is associated with three binary systems of the peritectic type. Draw the space diagram, and develop isotherms sufficient in number to display the internal structure of the diagram.

**3.** With reference to the space diagram of Fig. 14-1, draw isopleths of constant $B$ content at 0, 10, 20, 30, 40, 50, and 90% $B$.

**4.** Deduce the changes that should occur and the microstructure that should result from the cooling of the following alloys of the ternary eutectoid system illustrated in Fig. 14-12 from the temperature range in which only the $\gamma$ phase is present down to ordinary temperature: (*a*) the ternary eutectoid alloy, (*b*) gross composition selected at random within the range of the ternary eutectoid triangle, (*c*) gross composition somewhere upon the valley leading from the $CB$ eutectic to the ternary eutectoid point, (*d*) gross composition somewhere upon that line in the ternary eutectoid plane connecting $\beta$ with $\gamma$.

**5.** Deduce the course of freezing and the resulting microstructure of an alloy composed of equal parts of $A$, $B$, and $C$ in the ternary monotectic system of Fig. 14-13.

# CHAPTER 15

# TERNARY FOUR-PHASE EQUILIBRIUM—CLASS II

There is another type of ternary four-phase equilibrium, known as *four-phase equilibrium of the second kind* or *class* II *four-phase equilibrium,* that has no direct equivalent in binary systems. It may be thought of as being intermediate between eutectic and peritectic reaction:

$$L + \alpha \underset{\text{heating}}{\overset{\text{cooling}}{\rightleftharpoons}} \beta + \gamma$$

During heating or cooling, two phases interact to form two new phases.

The manner of representation of class II four-phase equilibrium in a space diagram is illustrated in Fig. 15-1, and isothermal sections through

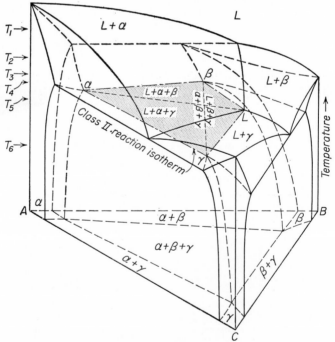

FIG. 15-1. Temperature-concentration diagram illustrating class II four-phase equilibrium.

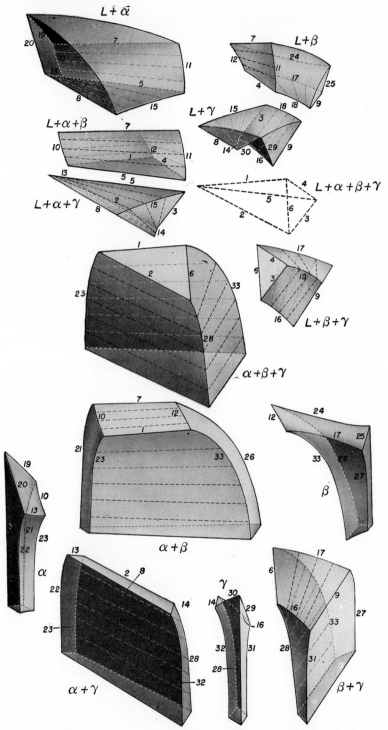

FIG. 15-2. Exploded model of the phase diagram of Fig. 15-1. The inscribed numbers identify edges that are identical lines on different segments of the model.

this diagram are given in Fig. 15-3. Further details are shown in the exploded model in Fig. 15-2. In these drawings it can be seen that *two* three-phase regions $L + \alpha + \beta$ and $L + \alpha + \gamma$ descend from higher temperature toward the four-phase reaction plane, where they meet to form a horizontal trapezium (kite-shaped figure) at the four corners of

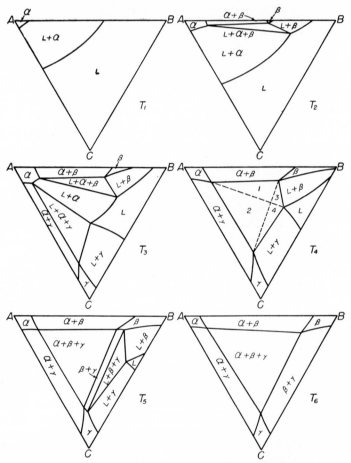

FIG. 15-3. Isotherms through the space diagram of Fig. 15-1.

which the four phases in equilibrium are represented: $\alpha + \beta + \gamma + L$. Beneath the four-phase reaction plane two new three-phase regions, $\alpha + \beta + \gamma$ and $L + \beta + \gamma$, originate and descend to lower temperature. Thus, the two tie-triangles, $L + \alpha + \beta$ and $L + \alpha + \gamma$, Fig. 15-4, join along a common tie-line $\alpha L$ to form the four-phase reaction trapezium, and this figure divides along the tie-line $\beta\gamma$ to form two new tie-triangles, $\alpha + \beta + \gamma$ and $L + \beta + \gamma$. An alloy occurring at the composition at which these tie-lines cross (intersection of the dashed lines in the central sketch of Fig. 15-4) would be composed entirely of $L$ and $\alpha$ just above the reaction temperature and entirely of $\beta$ and $\gamma$ just below the reaction temperature.

This construction conforms with the demands of the phase rule. Four-phase reaction should be univariant; the temperature and compositions of the participating phases should be constant at fixed pressure. These conditions are met, since the four-phase trapezium is horizontal (isothermal) and the one-phase regions touch it only at its four corners where a unique composition is thereby designated for each of the four phases.

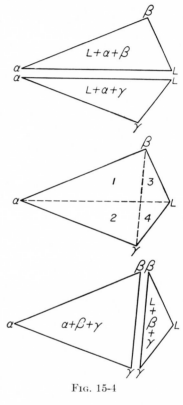

Other than the four-phase reaction trapezium, this diagram involves no regions that have not been discussed in previous chapters. The one-, two-, and three-phase regions in this diagram are, in fact, identical in number and designation with those found in the ternary eutectic diagram.

## Course of Freezing of Some Typical Alloys

An examination of the course of freezing of several typical alloys should prove helpful in gaining an understanding of the nature of class II four-phase reaction. Four alloys designated 1, 2, 3, and 4 in the central drawing of Fig. 15-4 will be con-

FIG. 15-4

sidered. It can be seen at once that these will undergo the following changes upon descending through the four-phase reaction temperature:

1. $L + \alpha + \beta \rightarrow \alpha + \beta + \gamma$
2. $L + \alpha + \gamma \rightarrow \alpha + \beta + \gamma$
3. $L + \alpha + \beta \rightarrow L + \beta + \gamma$
4. $L + \alpha + \gamma \rightarrow L + \beta + \gamma$

By reference to Fig. 15-3 it can be seen that alloy 1 begins freezing with a primary separation of the $\alpha$ phase followed by a secondary deposition of $\alpha + \beta$. Just above the four-phase reaction temperature the relative proportions of the three phases may be computed from the tie-triangle in solid lines in Fig. 15-5:

$$\%\beta = \frac{X1}{X\beta} \times 100 \approx 50\%$$

$$\%\alpha = \frac{XL}{\alpha L} \frac{1\beta}{X\beta} \times 100 \approx 28\%$$

$$\%L = \frac{\alpha X}{\alpha L} \frac{1\beta}{X\beta} \times 100 \approx 22\%$$

Just below the four-phase reaction temperature

$$\%\beta = \frac{Y1}{Y\beta} \times 100 \approx 66\%$$

$$\%\alpha = \frac{Y\gamma}{\alpha\gamma} \frac{1\beta}{Y\beta} \times 100 \approx 18\%$$

$$\%\gamma = \frac{\alpha Y}{\alpha\gamma} \frac{1\beta}{Y\beta} \times 100 \approx 16\%$$

The percentage of $\alpha$ has decreased sharply, and the liquid has disappeared altogether, while the quantity of $\beta$ has increased, and a substantial quantity of $\gamma$ has appeared as a new phase. Some of the previously existing $\alpha$ has been consumed by reaction with liquid, as in *peritectic reaction*. The increase in the quantity of $\beta$, coincident with the appearance of $\gamma$, more nearly resembles *eutectic reaction*. This is what was meant by the opening statement to the effect that class II four-phase equilibrium occupies a position midway between the eutectic and peritectic types.

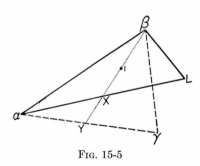

Fig. 15-5

Alloy 2 (Fig. 15-4) also begins freezing with a primary separation of the $\alpha$ phase, but the secondary crystallization will be of $\alpha + \gamma$ in this case instead of $\alpha + \beta$. If the tie-triangles were analyzed for this example, it would be found that at the four-phase reaction temperature, the liquid phase is wholly consumed and the $\alpha$ phase is partly redissolved to form a tertiary precipitation of $\beta + \gamma$.

The two alloys 3 and 4 which lie closer to the liquid corner of the four-phase trapezium differ from the foregoing pair of alloys in that the primary and secondary $\alpha$ crystals should be totally consumed if four-phase reaction goes to completion, while some of the liquid phase should remain to freeze as $\beta + \gamma$ at a lower temperature. In both alloys the primary constituent is $\alpha$, although its quantity may be small because these compositions are close to the lower edge of the $L + \alpha$ field. The secondary constituent in alloy 3 will be $\alpha + \beta$, while that in alloy 4 will be $\alpha + \gamma$. Both will suffer a loss of $\alpha$ and of some liquid to form $\beta + \gamma$ in four-phase reaction, and the simultaneous crystallization of $\beta$ and $\gamma$ will continue at lower temperature until the supply of liquid is exhausted.

As with peritectic alloys the formation of reaction layers (envelopment) upon the $\alpha$ phase may be expected to interfere with the completion of reaction by hindering the diffusion that must take place to establish equilibrium. Consequently, an excess of the $\alpha$ phase is likely to persist and some liquid is likely to survive through four-phase reaction in all the alloys of this system. This will have the effect of minimizing the differences in structure among the several alloys.

Some examples of characteristic microstructures of lead-antimony-tin alloys that undergo class II four-phase reaction are presented in Fig. 15-6. Isothermal sections of the appropriate portion of the lead-antimony-tin phase diagram are given in Fig. 15-7. The compositions at which the four alloys represented in Fig. 15-6 occur are designated $a$, $b$, $c$, and $d$ in the isotherm at 250°C in Fig. 15-7. Because no single etching technique serves to differentiate clearly among the $\beta$, $\delta$, and $\gamma$ phases of this alloy system, two pictures of each alloy are presented. The picture on the left, in each case, has been made by the use of an etch which blackens the $\delta$ phase and leaves the $\beta$ and $\gamma$ phases white, while the etch used in preparing the pictures on the right causes the $\delta$ phase to appear black, the $\gamma$ phase gray to black, and the $\beta$ phase white. Thus the $\beta$ phase is white and the $\delta$ phase black in both pictures, while the $\gamma$ phase is white in the pictures on the left and gray or black in the pictures on the right.

The large white squares are, in all cases, primary $\beta$. That they have been subject to "peritectic attack," especially in alloys $b$ and $d$ (pictures D and H), is evident from the penetration of the $\gamma$ phase along their edges. In alloy $c$ (pictures E and F) the primary constituent is $\gamma$, seen as typical dendrites, white in the picture on the left and black in the picture on the right. A two-phase constituent $\beta + \gamma$ forms secondarily in alloys $a$ and $c$ (pictures A, B, E, and F). The $\beta$ of this constituent is largely destroyed during subsequent transformation and remains only as small white irregular specks in the background of the microstructure, being more plentiful in alloy $a$ than in alloy $c$ as should be expected. The secondary constituent in alloys $b$ and $d$ (pictures C, D, G, and H) is $\beta + \delta$. Here again, the secondary $\beta$ is largely destroyed in subsequent reaction, being more plentiful in alloy $b$ than in alloy $d$. Four-phase reaction, in all alloys, proceeds by the formation of finely divided $\delta + \gamma$ through reaction of liquid with $\beta$ and, subsequently, by the direct decomposition of any remaining liquid into $\delta + \gamma$. This constituent appears as a mixture of very small black and white specks in the photographs on the left and of gray

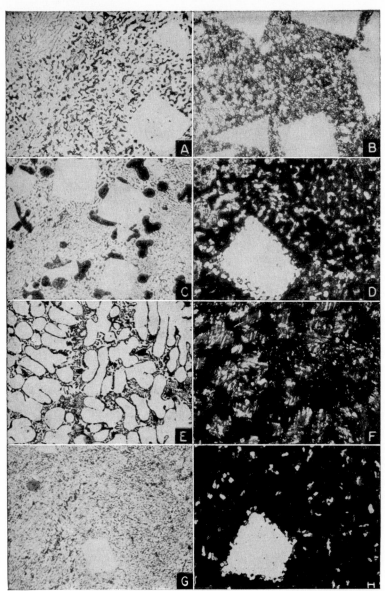

FIG. 15-6. Microstructure of cast alloys of compositions $a$, $b$, $c$, and $d$, designated in the first isotherm of Fig. 15-7. Two photomicrographs are required to show the structure of each alloy, because the $\gamma$ phase is indistinguishable from the $\delta$ phase (black) in the pictures on the right and from the $\beta$ phase (white) in the pictures on the left. In $a$, $b$, and $d$ (pictures A, B, C, D, G, and H) the primary constituent is $\beta$ which is partly redissolved in subsequent class II reaction, as shown by the serrated edges of the white squares. The primary constituent in alloy $c$ (pictures E and F) is dendritic $\gamma$. Tin-rich alloys $a$ and $c$ (pictures A, B, E, and F) display a secondary crystallization of $\beta + \gamma$, while lead-rich alloys $b$ and $d$ (pictures C, D, G, and H) have an equivalent $\beta + \delta$ constituent. In all cases, class II reaction results in the formation of a finely divided tertiary constituent of $\gamma + \delta$. Magnification: A and B, 500; C, 250; D, 750; E and F, 250; G, 250 and H, 750.

and black specks, which can scarcely be differentiated, in the photographs on the right.

## Other Examples of Class II Four-phase Equilibrium

Class II four-phase reaction appears to be relatively common in ternary alloy systems. Most of the examples on record are of the kind just described, in which one liquid and three solid phases participate. Similar reaction among four solid phases has been reported, however, in a

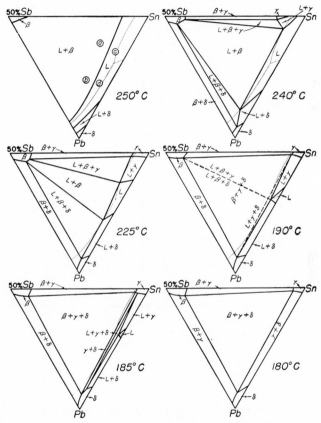

FIG. 15-7. Isotherms of the ternary system Pb-Sn-β (50% Sb + 50% Sn). The circled letters in the isotherm at 250°C indicate the compositions of the four alloys of which the microstructure is shown in Fig. 15-6.

number of systems, as have other combinations of phases in class II reaction. All conceivable combinations are listed in Table 4, Chap. 18, relatively few of which have as yet been observed in ternary alloy systems. Wherever such reaction does occur, however, the general principles outlined above should apply to its interpretation.

## Some Vertical Sections

In vertical sections through the space diagram the class II four-phase reaction plane appears, of course, as a horizontal line (see Figs. 15-8, 15-9, and 15-10). Otherwise, there is little generalization possible in the construction of vertical sections. A section $UV$ taken parallel to the $AB$

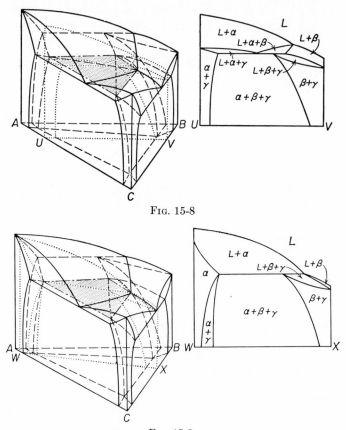

Fig. 15-8

Fig. 15-9

side of the space model is represented in Fig. 15-8. By shifting the plane of this section slightly so that it passes through the $\alpha$ and liquid compositions of the four-phase reaction, section $WX$ of Fig. 15-9 is obtained. Here the $L + \alpha$ field rests upon the four-phase reaction line because $\alpha L$ is an element of the trapezium and also the lowest tie-line in the $L + \alpha$ field. Section $YZ$ of Fig. 15-10 is taken parallel to the $BC$ side of the diagram and illustrates the complexity of configuration that is sometimes encountered in vertical sections through this type of ternary diagram. Isothermal sections, Fig. 15-3, are much easier to understand in this case.

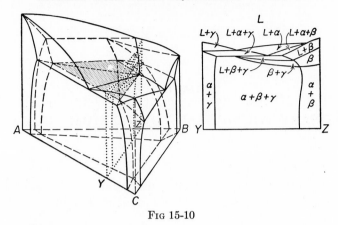

FIG 15-10

## Heat Treatment of Alloys

There is little to be said concerning the relationships between the constitution of alloys of class II ternary four-phase systems and the structures produced by heat treatment that has not already been said in connection with the ternary eutectic system. The discussion with regard to Fig. 14-11 applies as well to alloys in the neighborhood of the $A$ component in the diagram just considered. Similar relationships obtain in other alloys of this type of system, the most noteworthy difference being that liquation can occur below the temperature of four-phase reaction in any alloys containing both the $\beta$ and the $\gamma$ phase. These phases will melt together at the $BC$ binary eutectic temperature (or somewhat above) if one or both are not dissolved prior to the attainment of this temperature.

### PRACTICE PROBLEMS

**1.** How do the percentages of the several phases change during the class II transformation of composition 3 in Fig. 15-4? Of composition 4? If Class II transformation were to be suppressed altogether, as by rapid cooling or a very slow diffusion rate, how would the microstructures of these alloys (at room temperature) differ from that produced by complete transformation?

**2.** Draw the space diagram of a ternary system made up of two binary peritectic-type systems and one eutectic-type system and having a class II four-phase equilibrium below the two binary peritectic temperatures but above the temperature of the binary eutectic. Develop isothermal sections sufficient to display the internal structure of this diagram.

# CHAPTER 16

# TERNARY FOUR-PHASE EQUILIBRIUM—CLASS III

The inverse of class I four-phase equilibrium is *class* III *four-phase equilibrium*, the ternary equivalent of peritectic equilibrium:

$$L + \alpha + \beta \underset{\text{heating}}{\overset{\text{cooling}}{\rightleftharpoons}} \gamma$$

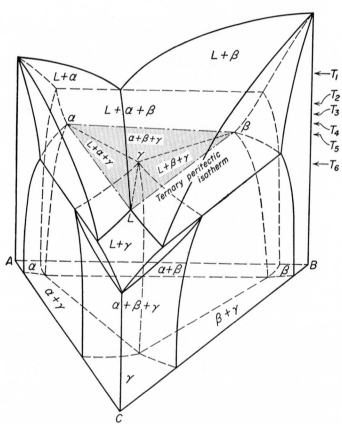

FIG. 16-1. Temperature-composition diagram of an ideal ternary peritectic system, class III.

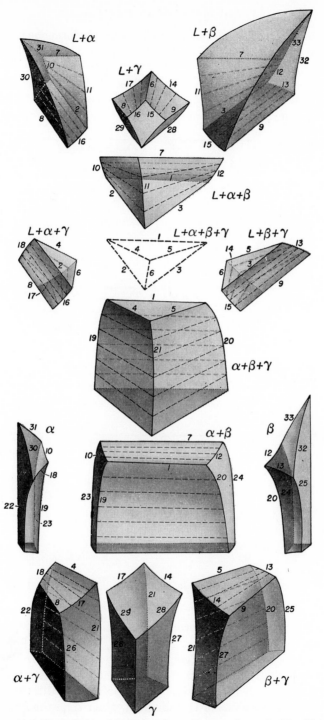

FIG. 16-2. Exploded model of the phase diagram shown in Fig. 16-1.

187

Three phases interact isothermally upon cooling to form one new phase, or conversely, upon heating, one phase decomposes isothermally into three new phases.

Class III four-phase equilibrium is illustrated in the diagram of Fig. 16-1. Three phases $\alpha$, $\beta$, and $L$, located at the corners of a horizontal triangular reaction plane, combine to form the $\gamma$ phase, whose composition

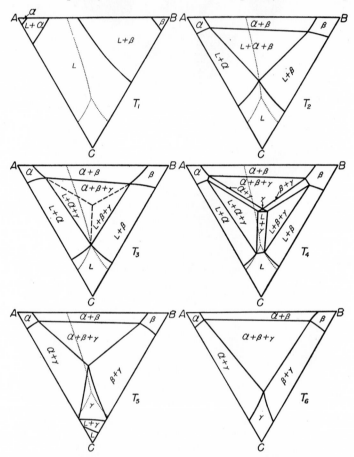

FIG. 16-3. Isotherms through the space diagram of Fig. 16-1.

lies within the triangle. One three-phase field $L + \alpha + \beta$ descends from higher temperature to the ternary peritectic temperature, and three three-phase regions $\alpha + \beta + \gamma$, $L + \alpha + \gamma$, and $L + \beta + \gamma$ issue beneath and proceed to lower temperature (see Fig. 16-3). The obvious equivalence of this construction to that of the ternary eutectic disposes of any necessity of further demonstrating its conformity with the phase rule.

Once again, the regions of this diagram, other than the four-phase reaction plane itself, are similar in designation and general form to those found in the two preceding chapters (see Fig. 16-2). Edges along which the various regions meet in the assembled diagram have been identically numbered in order to facilitate the visualization of the diagram.

## Freezing of Alloys

Although class III reaction is probably fairly common among ternary alloy systems, no example has yet been studied in detail, and it is therefore not feasible to support an analysis of ternary peritectic behavior with experimental observation. The perfection of the correlation between constitution and reaction behavior in all the examples discussed up to this point is such, however, as to lend confidence in the reliability of the deductive approach. The analysis that follows is wholly deductive.

Three essentially distinct types of class III four-phase reaction should occur during cooling:

1. $L + \alpha + \beta \rightarrow \alpha + \beta + \gamma$
2. $L + \alpha + \beta \rightarrow L + \alpha + \gamma$
3. $L + \alpha + \beta \rightarrow L + \beta + \gamma$

corresponding to alloys falling within the zones numbered 1, 2, and 3 in Fig. 16-4. In each case, the phases present above the ternary peritectic temperature all either diminish in quantity or disappear altogether. This

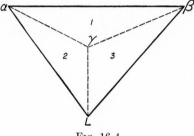

Fig. 16-4

can be demonstrated by analyzing the tie-triangles by the use of the lever principle. Consider alloy $X$, Fig. 16-5, an alloy of the first type. Above the peritectic temperature the relative proportions of the three phases should be

$$\%\beta = \frac{WX}{W\beta} \times 100 \approx 42\%$$

$$\%\alpha = \frac{WL}{\alpha L} \frac{X\beta}{W\beta} \times 100 \approx 42\%$$

$$\%L = \frac{\alpha W}{\alpha L} \frac{X\beta}{W\beta} \times 100 \approx 16\%$$

and just below this temperature

$$\%\beta = \frac{YX}{Y\beta} \times 100 \approx 34\%$$

$$\%\alpha = \frac{Y\gamma}{\alpha\gamma} \frac{X\beta}{Y\beta} \times 100 \approx 33\%$$

$$\%\gamma = \frac{\alpha Y}{\alpha\gamma} \frac{X\beta}{Y\beta} \times 100 \approx 33\%$$

The liquid phase has vanished and the quantities of $\alpha$ and $\beta$ are sharply reduced. A similar condition obtains with alloys of the second and third types.

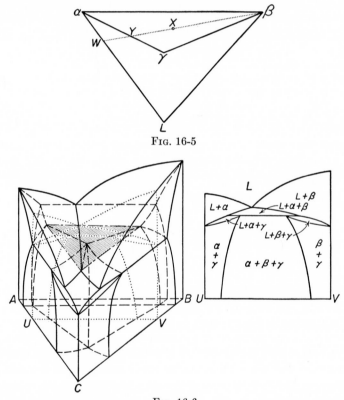

FIG. 16-5

FIG. 16-6

Such reaction requires that the $\gamma$ phase be formed preferentially at three-way junctions of $L$, $\alpha$, and $\beta$ in the structure of the alloy, because the composition of the $\gamma$ phase requires the borrowing of $A$ component from the $\alpha$, $B$ component from the $\beta$, and $C$ component from the liquid. As with binary peritectic reaction, it is to be expected that the formation of the γ phase should retard its own further growth by lengthening the path over which solid-phase diffusion must act to supply the necessary components. Thus, incomplete reaction should be common; any remaining liquid would, of course, freeze to cored γ.

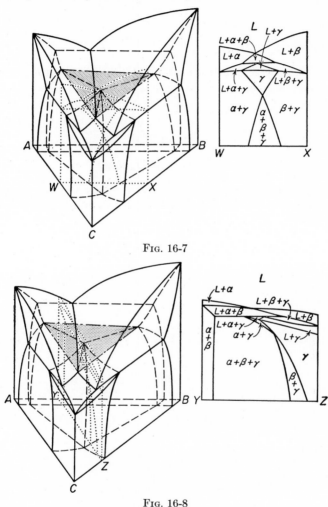

Fig. 16-7

Fig. 16-8

## Occurrence of Class III Four-phase Equilibrium

No example of class III equilibrium corresponding to the diagram discussed above, in which the low-temperature phase is one of the terminal solid solutions, has been reported in the research literature. There are, however, examples of the interaction of liquid and three solid phases in which the low-temperature solid is an intermediate phase ("ternary compound"). A tabulation of all conceivable combinations of phases in class III four-phase equilibrium is given in Table 4, Chap. 18. Future constitutional studies will probably reveal the existence of a number of these.

## Vertical Sections

Three typical vertical sections through the space diagram are presented in Figs. 16-6, 16-7, and 16-8. As should be expected, the four-phase reaction plane is intersected in a horizontal line in all sections. Here again the complexity of the vertical sections, as well as their other limitations, make the use of isothermal sections preferable.

### PRACTICE PROBLEM

1. Deduce the cast microstructures of alloys 1, 2, and 3 in Fig. 16-4, assuming, first, that equilibrium is maintained during cooling and, second, that natural freezing conditions prevail.

CHAPTER 17

# CONGRUENT TRANSFORMATION IN TERNARY SYSTEMS

Congruently melting phases in ternary alloy systems are of three kinds, namely, (1) the pure components, (2) binary intermediate phases of congruent characteristic, and (3) ternary intermediate phases that melt and transform without composition change. Intermediate ternary phases may be either congruently or incongruently melting. The latter melt by class III four-phase reaction, but the former behave as pure substances and may, at times, be considered as components of an alloy system. Many of the ternary intermediate phases have compositions approximating simple ratios of the three kinds of atoms concerned and are often referred to as *ternary compounds*, but as with binary alloys, the use of the term *compound* in this connection has been in disfavor. No sharp distinction is recognized at present between phases that occur with simple proportions of the three components and those that do not, so that the more general term *ternary intermediate phase* is preferred.

## Quasi-binary Systems

When a congruently melting intermediate phase occurs in a ternary system, it sometimes happens that this phase forms a *quasi-binary system* with one of the other components. The example given in Fig. 17-1 shows a quasi-binary system existing between the intermediate $\delta$ phase of the binary system $AB$ and the $C$ component of the ternary system. A vertical section taken along the line joining these two compositions will be, in all respects, equivalent to a binary phase diagram (Fig. 17-2). All tie-lines in two-phase fields will lie in the plane of the section, and three-phase equilibrium will be represented by a single horizontal line. No four-phase equilibrium can occur in the quasi-binary section. All alloys behave as though binary in their structural response to temperature change.

The quasi-binary section divides the ternary diagram into two independent portions just as a congruently melting and transforming binary phase divides the binary diagram into independent parts. This condition is more clearly evident in the isothermal sections of Fig. 17-3 than in the space diagram itself. A dotted line has been drawn between the $C$ corner

193

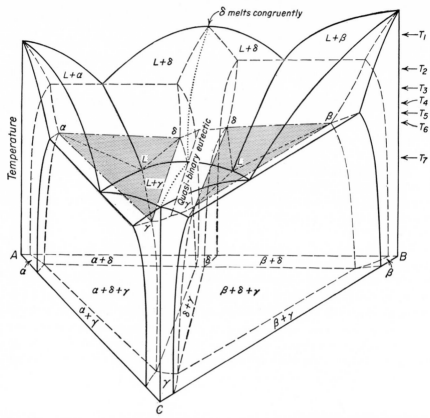

FIG. 17-1. Temperature-composition diagram of a ternary alloy system displaying a quasi-binary section in the vertical plane between the $C$ component and the congruently melting binary $\delta$ phase.

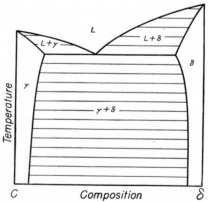

FIG. 17-2. A quasi-binary section from Fig. 17-1 has tie-lines in the plane of the section in all two-phase regions.

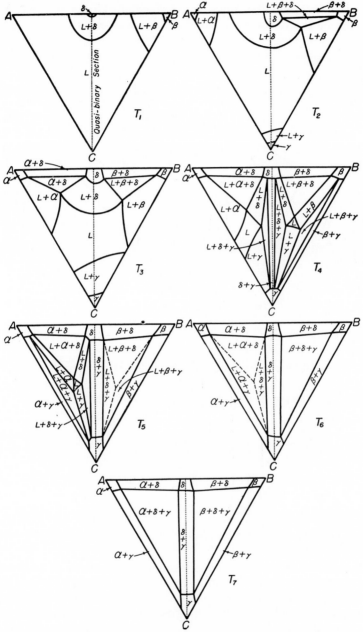

FIG. 17-3. Isotherms through the space diagram of Fig. 17-1.

195

of each isotherm and the opposite intermediate binary phase $\delta$. Upon either side of the dotted line the complete configuration of a ternary eutectic diagram, such as that illustrated in Fig. 14-3, is found.

In the isotherm at $T_4$, Fig. 17-3, it can be seen that there is a region designated $L + \gamma + \delta$ on each side of the quasi-binary section. Both of these regions originate upon the quasi-binary eutectic isotherm

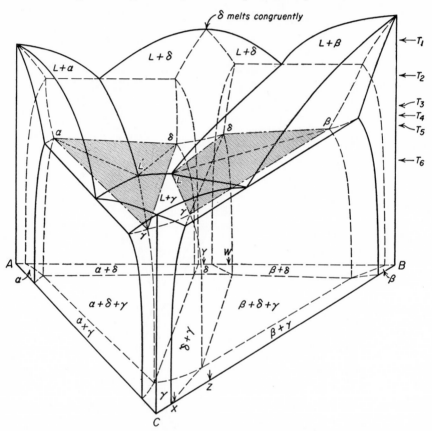

Fig. 17-4. Temperature-composition space diagram having a congruently melting intermediate phase but no quasi-binary section.

$(L \rightarrow \gamma + \delta)$, which occurs at the maximum temperature of equilibrium among these three phases. It will be recalled from Chap. 13 that a three-phase region, passing through a temperature maximum, is reduced to a single line at the maximum. The liquidus surface upon the two sides of the quasi-binary section forms a saddle astride the quasi-binary eutectic composition; see the liquidus projection of this diagram in Fig. 17-8a. It would be possible to have the quasi-binary eutectic correspond to a minimum in the ternary liquidus surface if the $L + \gamma + \delta$ regions passed through a minimum instead of a maximum as in Fig. 13-31. This could occur if class II four-phase reactions were substituted for the class I reactions shown in Fig. 17-1. The existence of a congruently melting intermediate phase in a ternary system by no means guarantees that a quasibinary system will exist. A contrary case involving the same phases in the

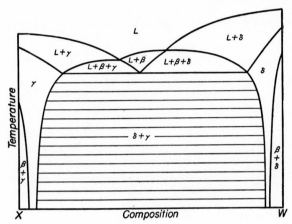

FIG. 17-5. Isopleth from the space diagram of Fig. 17-4.

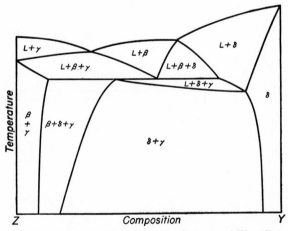

FIG. 17-6. Isopleth from the space diagram of Fig. 17-4.

same binary configuration appears in Fig. 17-4. Here there is no quasibinary section (Fig. 17-7). A class II four-phase reaction isotherm $L + \beta + \gamma + \delta$ crosses the isopleth between the $\gamma$ and $\delta$ phases (see $T_4$ in Fig. 17-7). Tie-lines are found in the isopleth of Fig. 17-5 only in the $\gamma + \delta$ field. Those of the $L + \beta$, $L + \gamma$, and $L + \delta$ fields lie at an angle to the plane of the section. Three-phase equilibrium is represented by areas, i.e., the areas $L + \beta + \gamma$ and $L + \beta + \delta$. Alloys of this series are not set apart from those on either side of the vertical section by their binary behavior but behave as ternary alloys.

The diagram of Fig. 17-5 has, as a matter of fact, been drawn in the most favorable way to place as many as possible of the tie-lines in the isopleth. It has been assumed in making the drawing that the $\gamma\delta$ tie-line

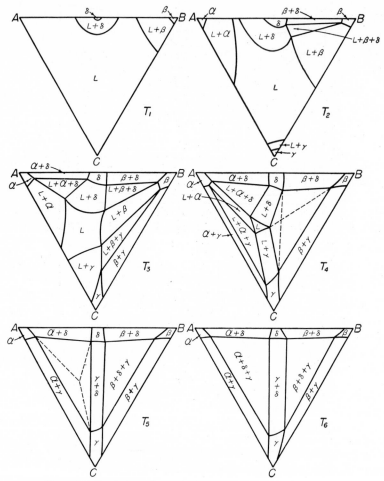

Fig. 17-7. Isotherms through the space diagram of Fig. 17-4.

of the class II four-phase reaction plane lies exactly in the vertical section. It is not necessary that this condition obtain. If the vertical section crosses the four-phase isotherm at random, the isopleth might appear somewhat as in Fig. 17-6, where no tie-lines exist anywhere in the section. The relative likelihood of tie-lines running parallel to the $\gamma\delta$ section in the several two-phase fields can be ascertained most easily by reference to the isotherms of Fig. 17-7. Here it can be seen that tie-lines in the $L + \beta$ field will lie almost perpendicular to the section while those in the other two-phase fields lie at various angles. Certain tie-lines in the $L + \gamma$, $L + \delta$, and $\gamma + \delta$ fields might, by chance, lie in the section. No maximum or minimum of three-phase equilibrium occurs in the vertical section; the liquidus valleys pass this vertical plane without inflection as can be seen in Fig. 17-8b.

Quasi-binary equilibrium can exist also between congruently melting ternary intermediate phases and other components (Fig. 17-9). One ternary intermediate phase is shown forming three quasi-binary systems,

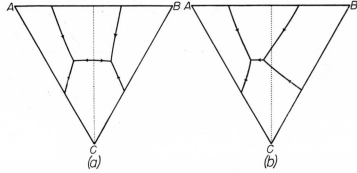

Fig. 17-8. Liquidus valleys from the two foregoing space diagrams: (a) Fig. 17-1 and (b) Fig. 17-4.

one with each of the terminal components. This divides the space diagram into three independent sections; three separate ternary eutectic systems are shown (see also Fig. 17-10). These three quasi-binary systems are identical in all characteristics with those formed by binary intermediate phases.

## Division of the Ternary Diagram

A single quasi-binary system divides the ternary system into two parts. Two quasi-binary systems divide the ternary into three parts, three quasi binaries divide the ternary into four parts, and so on (see Fig. 17-11). The maximum number of quasi-binary sections is equal to the number of congruent binary intermediate phases. This is true regardless of the manner in which the sections are selected; for example, compare drawings $b$ and $c$ or $d$, $e$, and $f$ in Fig. 17-11. The maximum number of independent ternary systems $n$ into which the main ternary diagram is thus divided is equal to the number of binary intermediate phases $b$ plus 1:

$$n = b + 1$$

By reference to Fig. 17-12a, b, and c, it can be seen that one ternary inter-mediate phase can give rise to three quasi-binary systems and that the maximum number of independent ternary systems into which the diagram is thus divided is equal to twice the number of ternary intermediate phases $t$ plus 1:

$$n = 2t + 1$$

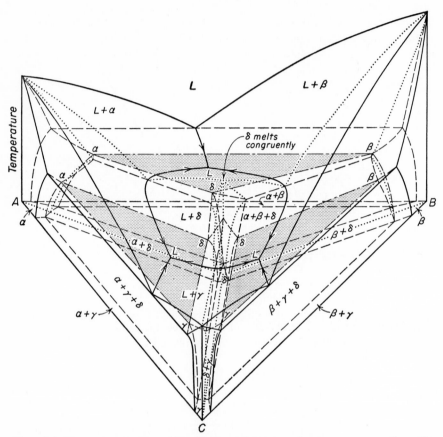

Fig. 17-9. Temperature-composition diagram of a system having a congruently melt-ing ternary intermediate phase that forms quasi-binary sections with each of the three components.

Where binary and ternary intermediate phases occur in the same ternary system,

$$n = b + 2t + 1$$

In the drawing of Fig. 17-12f there are two ternary and one binary inter-mediate phases:

$$n = 1 + 2 \times 2 + 1 = 6$$

This rule is of interest in the construction of ternary diagrams but finds its greatest usefulness in the checking of isotherms. Here it is unnecessary to consider whether or not the intermediate phases are congruently melting, for the conditions at the temperature of the isotherm alone are concerned. Let $n$ be the number of three-phase tie-triangles in the isotherm,

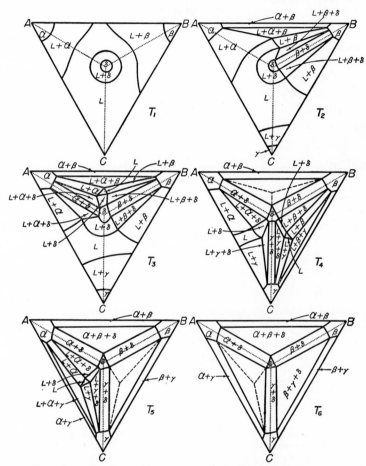

FIG. 17-10. Isotherms through the space diagram of Fig. 17-9.

$b$ the number of binary phases (including all phases that touch the edge of the diagram, except the terminal phases), and $t$ the number of phases occurring wholly within the diagram. Then the number of three-phase regions will be equal to the number of binary phases plus twice the number of ternary phases plus 1. In applying this rule it is necessary to make an exception of phases which are isomorphous across the ternary diagram. For each isomorphous phase one three-phase region must be

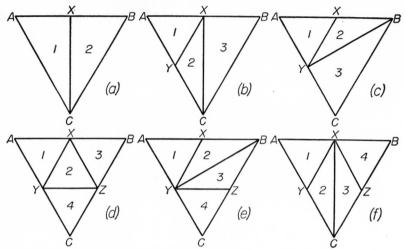

FIG. 17-11. Possible arrangement of quasi-binary sections in the presence of one (a), two (b and c), and three (d, e, and f) binary intermediate phases.

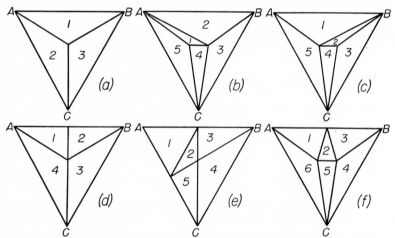

FIG. 17-12. Possible arrangement of quasi-binary sections in the presence of one (a) and two (b and c) ternary intermediate phases, and combinations of one binary and one ternary intermediate phase (d), two binary and one ternary intermediate phase (e), and one binary and two ternary intermediate phases (f).

subtracted from the result of the above computation. Where one phase is isomorphous with more than one other binary phase, the number of isomorphous systems is one less than the number of binary phases participating.

Thus, in section $T_4$ of Fig. 17-3 there are two binary phases ($\delta$ and $L$) and one ternary phase ($L$).

$$n = 2 + 2 \times 1 + 1 = 5$$

Five tie-triangles can be counted in the isotherm. In section $T_2$, Fig. 17-3, there are two binary ($\delta$ and $L$) and no ternary phases, but the $L$ phase is isomorphous in two limbs.

$$n = 2 + 0 + 1 - 2 = 1$$

Only one three-phase region is found in this section.

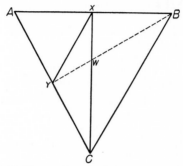

FIG. 17-13. Illustrating the "clear-cross principle."

## The "Clear-cross Principle"

It has been shown in Figs. 17-11 and 17-12 that with a given number of intermediate phases, the quasi-binary sections can be arranged in several different ways. This is illustrated in Fig. 17-13, where the quasi-binary sections may be either $XY$ and $XC$ or $XY$ and $YB$. Obviously, both $XC$ and $YB$ cannot exist; at least one of these two must be false.

An experimental procedure, known as the "clear-cross principle," has been devised to ascertain which of two quasi-binary systems (if either) is real. An alloy of composition $W$, lying on the intersection of the two potential quasi-binary sections, is made. After homogenization to establish equilibrium, the alloy is examined to determine the phases present. If only $X$ and $C$ are found, then the quasi-binary section $XC$ is real and the section $YB$ is false. If phases $Y$ and $B$ alone are found, then section $YB$ is real and $XC$ is false. But if three phases are found, neither is real.

## Some Other Examples of Congruency

Congruent melting of ternary isomorphous solid solutions at compositions of maximum or minimum melting temperature has been mentioned in Chap. 12. A corresponding congruent transformation within the solid state is also possible. This may involve an order-disorder type of transformation as explained in Chap. 7. Likewise, quasi-binary behavior extends to equilibria wholly within the solid state.

# CHAPTER 18

# COMPLEX TERNARY SYSTEMS

Having examined the structural units from which ternary phase diagrams are built, it remains to consider the ways in which these may be assembled into the relatively complex diagrams that are characteristic of the majority of ternary alloy systems. Portions of many three-component metal systems have been examined in the laboratory, and phase diagrams have been published, but in very few cases can it be said that the ternary diagram has been even approximately established in all its parts. Indeed, only small areas of the diagram adjacent to alloy compositions of immediate technical interest have been investigated in the majority of ternary systems. This is a pronounced handicap in the interpretation of such diagrams, where, as often happens, the composition ranges of some of the phases which participate in reactions in the explored areas themselves lie in unexplored areas. For this reason it is of more than academic interest to be able to speculate intelligently concerning constructions that are possible in regions of diagrams relative to which only fragmentary information is available. Practice in the synthesis of ternary diagrams from real or assumed data is useful in developing skill in this type of reasoning.

## Review of Structural Units

Before proceeding with the assembly of whole diagrams, the structural units that have been discussed in preceding chapters will be summarized in review (see Fig. 18-1). *One-phase equilibria* are represented by spaces of any shape, bounded by surfaces of the two-phase regions and, hence, cannot be represented by any characteristic structural unit.

*Two-phase equilibrium* is always represented by spaces having two conjugate surfaces that are joined at every point by tie-lines connecting the compositions of the conjugate phases (Fig. 18-1a). Where congruent transformation occurs between the two phases, the bounding surfaces of the two-phase region meet at a single point as at the melting point of a pure component (Fig. 18-1b) or at the congruent melting point of an

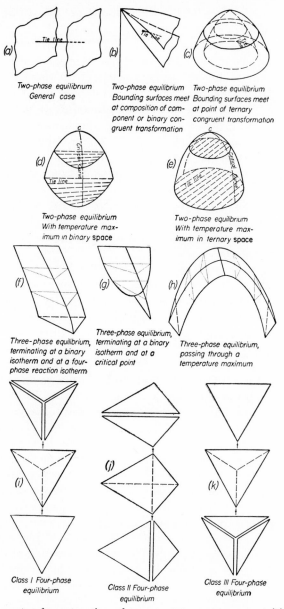

(a) Two-phase equilibrium General case

(b) Two-phase equilibrium Bounding surfaces meet at composition of component or binary congruent transformation

(c) Two-phase equilibrium Bounding surfaces meet at point of ternary congruent transformation

(d) Two-phase equilibrium With temperature maximum in binary space

(e) Two-phase equilibrium With temperature maximum in ternary space

(f) Three-phase equilibrium, terminating at a binary isotherm and at a four-phase reaction isotherm

(g) Three-phase equilibrium, terminating at a binary isotherm and at a critical point

(h) Three-phase equilibrium, passing through a temperature maximum

(i) Class I Four-phase equilibrium

(j) Class II Four-phase equilibrium

(k) Class III Four-phase equilibrium

FIG. 18-1. Elements of construction of ternary temperature-composition diagrams.

205

intermediate phase (Fig. 18-1c). Such points are always at temperature maxima or minima.

A region of two-phase equilibrium may also terminate in a critical point which may lie in one of the binary faces of the space diagram (Fig. 18-1d) or may lie within the ternary space of the diagram (Fig. 18-1e). The critical point itself represents the highest or lowest temperature at which the two phases may coexist and is, therefore, a maximum or minimum point. There is also a critical curve, which connects the points on successive isotherms at which the tie-lines are reduced to zero length.

*Three-phase equilibrium* is represented by a trio of conjugate curves everywhere connected by tie-triangles. Such a region may terminate in a horizontal line at a temperature maximum or minimum at a binary three-phase isotherm (Fig. 18-1f) or in space (Fig. 18-1g) or at a quasi-binary section (Fig. 18-1h). It may also terminate in a final tie-triangle at a temperature minimum or maximum where a four-phase reaction plane is intersected (Fig. 18-1f).

*Four-phase equilibrium* may be represented by any of three distinct constructions. In class I four-phase equilibrium three three-phase regions involving a total of four phases approach a triangular reaction plane from higher temperature, one phase vanishes, and a single three-phase region descends to lower temperature (Fig. 18-1i). Eighteen combinations of four phases that may conceivably join in class I reaction are listed in Table 4. Among these, reaction Ih will be recognized as the ternary eutectic, Io as the ternary eutectoid, and Ik as a version of the ternary monotectic which was discussed at the end of Chap. 14. No names have been applied to any but the ternary eutectic and ternary eutectoid.

Class II equilibrium (Fig. 18-1j) occurs when four phases represented in two three-phase regions that have two phases in common approach from higher temperature and join along a common tie-line to form a trapezium at the four corners of which the compositions of the four phases are designated. Reaction is represented by the division of the trapezium into two tie-triangles having the opposite two phases in common, the corresponding three-phase regions proceeding to lower temperature. A list of 21 conceivable class II reactions is given in Table 4. Reaction IIp was employed as a typical example in Chap. 15. Several of the others are encountered in existing phase diagrams. None has a special name; it is not correct to refer to any class II reaction as a ternary peritectic reaction.

The third form of four-phase equilibrium, class III, is represented as in Fig. 18-1k. One three-phase region descends from higher temperature and splits into three three-phase regions isothermally at a triangular reaction plane where the fourth phase first makes its appearance. This is the peritectic type of ternary reaction. Eighteen combinations of phases that could conceivably partake in class III four-phase reaction are listed in Table 4. Reaction IIIo, in which the $\gamma$ phase is an intermediate that melts incongruently, is probably the most common in ternary phase diagrams. No special names have been assigned to any of these reactions, but the use of the name "ternary peritectic" for IIIo and "ternary peritectoid" for IIIr seems logical.

TABLE 4. TYPES OF TERNARY FOUR-PHASE EQUILIBRIA

| Class I | Class II | Class III |
|---|---|---|
| I$a$. $G \rightarrow L_I + L_{II} + L_{III}$ | II$a$. $G + L_I \rightarrow L_{II} + L_{III}$ | III$a$. $G + L_I + L_{II} \rightarrow L_{III}$ |
| I$b$. $G \rightarrow L_I + L_{II} + \alpha$ | II$b$. $G + L_I \rightarrow L_{II} + \alpha$ | III$b$. $G + L_I + L_{II} \rightarrow \alpha$ |
| I$c$. $G \rightarrow L_I + \alpha + \beta$ | II$c$. $G + L_I \rightarrow \alpha + \beta$ | III$c$. $G + L_I + \alpha \rightarrow L_{II}$ |
| I$d$. $G \rightarrow \alpha + \beta + \gamma$ | II$d$. $G + \alpha \rightarrow L_I + L_{II}$ | III$d$. $G + L_I + \alpha \rightarrow \beta$ |
| I$e$. $L_I \rightarrow G + L_{II} + L_{III}$ | II$e$. $G + \alpha \rightarrow L_I + \beta$ | III$e$. $G + \alpha + \beta \rightarrow L_I$ |
| I$f$. $L_I \rightarrow G + L_{II} + \alpha$ | II$f$. $G + \alpha \rightarrow \beta + \gamma$ | III$f$. $G + \alpha + \beta \rightarrow \gamma$ |
| I$g$. $L_I \rightarrow G + \alpha + \beta$ | II$g$. $L_I + L_{II} \rightarrow L_{III} + L_{IV}$ | III$g$. $L_I + L_{II} + L_{III} \rightarrow G$ |
| I$h$. $L_I \rightarrow \alpha + \beta + \gamma$ | II$h$. $L_I + L_{II} \rightarrow L_{III} + \alpha$ | III$h$. $L_I + L_{II} + L_{III} \rightarrow L_{IV}$ |
| I$i$. $L_I \rightarrow L_{II} + L_{III} + L_{IV}$ | II$i$. $L_I + L_{II} \rightarrow \alpha + \beta$ | III$i$. $L_I + L_{II} + L_{III} \rightarrow \alpha$ |
| I$j$. $L_I \rightarrow L_{II} + L_{III} + \alpha$ | II$j$. $L_I + L_{II} \rightarrow G + L_{III}$ | III$j$. $L_I + L_{II} + \alpha \rightarrow G$ |
| I$k$. $L_I \rightarrow L_{II} + \alpha + \beta$ | II$k$. $L_I + L_{II} \rightarrow G + \alpha$ | III$k$. $L_I + L_{II} + \alpha \rightarrow L_{III}$ |
| I$l$. $\alpha \rightarrow G + L_I + L_{II}$ | II$l$. $L_I + \alpha \rightarrow G + L_{II}$ | III$l$. $L_I + L_{II} + \alpha \rightarrow \beta$ |
| I$m$. $\alpha \rightarrow G + L_I + \beta$ | II$m$. $L_I + \alpha \rightarrow G + \beta$ | III$m$. $L_I + \alpha + \beta \rightarrow G$ |
| I$n$. $\alpha \rightarrow G + \beta + \gamma$ | II$n$. $L_I + \alpha \rightarrow L_{II} + L_{III}$ | III$n$. $L_I + \alpha + \beta \rightarrow L_{II}$ |
| I$o$. $\alpha \rightarrow \beta + \gamma + \delta$ | II$o$. $L_I + \alpha \rightarrow L_{II} + \beta$ | III$o$. $L_I + \alpha + \beta \rightarrow \gamma$ |
| I$p$. $\alpha \rightarrow L_I + L_{II} + L_{III}$ | II$p$. $L_I + \alpha \rightarrow \beta + \gamma$ | III$p$. $\alpha + \beta + \gamma \rightarrow G$ |
| I$q$. $\alpha \rightarrow L_I + L_{II} + \beta$ | II$q$. $\alpha + \beta \rightarrow G + L_I$ | III$q$. $\alpha + \beta + \gamma \rightarrow L_I$ |
| I$r$. $\alpha \rightarrow L_I + \alpha + \beta$ | II$r$. $\alpha + \beta \rightarrow G + \gamma$ | III$r$. $\alpha + \beta + \gamma \rightarrow \delta$ |
| | II$s$. $\alpha + \beta \rightarrow L_I + L_{II}$ | |
| | II$t$. $\alpha + \beta \rightarrow L_I + \gamma$ | |
| | II$u$. $\alpha + \beta \rightarrow \gamma + \delta$ | |

## Assembly of the Ternary Diagram

The assembly of the structural units into a complete diagram involving a variety of forms, such as the example of Fig. 18-2, is assisted by observing the following rules:

1. *One-phase* regions may meet each other only at single points, which are also temperature maxima or minima.

2. *One-phase* regions are elsewhere separated from each other by two-phase regions representing the two phases concerned; thus, the bounding surfaces of one-phase regions are always boundaries of two-phase regions.

3. *One-phase* fields touch three-phase regions only along lines which are generally nonisothermal.

4. *One-phase* regions touch four-phase reaction planes only at single points.

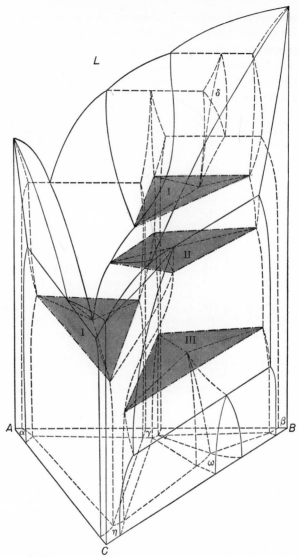

Fig. 18-2. A hypothetical ternary temperature-composition diagram involving all three classes of four-phase equilibrium.

5. *Two-phase* regions touch each other only along lines which, in general, are nonisothermal.

6. *Two-phase* regions are elsewhere separated by one- and three-phase regions by whose bounding surfaces they are enclosed.

7. *Two-phase* regions meet three-phase regions upon "ruled" bounding surfaces generated by the limiting tie-lines.

8. *Two-phase* regions touch four-phase reaction planes along single isothermal lines, which are limiting tie-lines.

9. *Three-phase* regions meet each other nowhere except at the four-phase reaction isotherms.

10. *Three-phase* regions are elsewhere separated and bounded by two-phase regions involving those two phases which are held in common by the neighboring three-phase regions.

These conditions can be summarized as follows: *A field of the phase diagram, representing equilibrium among a given number of phases, can be bounded only by regions representing equilibria among one more or one less than the given number of phases.* This statement is quite general; it applies

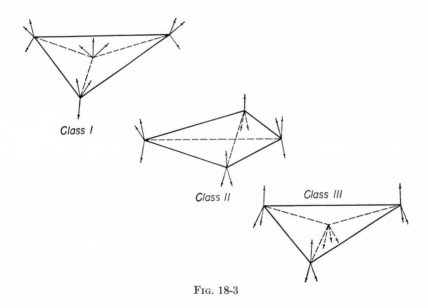

Fɪɢ. 18-3

to binary systems and to those of four and more components as well as to the ternary systems. In applying this rule it is admissible, and proper to consider the univariant equilibrium isotherms as "regions."

In constructing a diagram such as that shown in Fig. 18-2, it is well to remember that each four-phase isotherm must be associated with four three-phase regions, six two-phase regions, and four one-phase regions. This arrangement requires that three lines (no more and no less) shall meet the reaction isotherm at each of the four points at which it touches the one-phase regions. Twelve lines in all are attached to each four-phase isotherm (see Fig. 18-3). Two four-phase equilibria that have three phases in common may be connected by a three-phase region involving the three phases concerned. More than one three-phase region may never run between any pair of four-phase reaction isotherms unless they have all four phases in common.

Attention is directed again to the rule of construction illustrated in Fig. 13-7, which states that in the isothermal section through the ternary space diagram, the boundaries of a one-phase region where it meets a three-phase region must be so drawn that if projected, they would *both* extend into two-phase regions or *both* extend into the three-phase region but *never* one into a two-phase region while the other projects into the three-phase region or either or both into the one-phase region. The structure of isotherms may be further verified by the use of the rule, described at the end of the previous chapter, which relates the number of three-phase equilibria $n$ to the number of binary $b$ and ternary $t$ intermediate phases in the system $n = b + 2t + 1$.

The division of the ternary diagram into independent sections by quasi-binary sections greatly simplifies the structure of the diagram, because each section may then be treated separately. True quasi-binary sections are by no means so common, however, among "metal" systems as they are among "chemical" systems, where many very stable compounds exist. It is usually unwise to assume the existence of a quasi-binary system unless there is definite evidence of its reality.

### Ternary Models for Practice

At the end of this chapter a group of practice problems is given. The construction of the "problem diagrams" with due observance of the

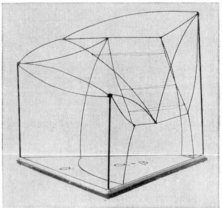

Fig. 18-4. Wire model of the space diagram illustrated in Fig. 13-18.

foregoing rules and generalizations should be of considerable assistance in developing a "feel for ternary diagrams." Many students of the subject find it helpful also to build models of the simpler ternary diagrams. This may be done easily by the use of artists' modeling clay. Each field is modeled separately in clay of a different color, and the various pieces are fitted into the completed diagram. When finished, the model may be cut horizontally or vertically to obtain the isotherms and isopleths. Another very satisfactory method of making a model of a ternary diagram is by the use of wires to represent all the lines in the diagram, (Fig. 18-4). Cardboard may be used to make the four-phase reaction planes, and colored threads strung horizontally around the wires bounding the three edges of each three-phase region will serve to produce tie-triangles. This kind of model cannot be sectioned, of course, but it has the advantage that the structure can be "seen through" and each field observed *in situ*. Regardless of the method used, it is advisable to make a practice of making sketches of isothermal sections through all space models studied, because this is the form in which the majority of real diagrams are presented. The ultimate objective of working with ternary models is to gain proficiency in the interpretation of isothermal sections.

### Interpretation of Complex Ternary Diagrams

Although not always easy to apply, the principles of the interpretation of complex ternary diagrams are themselves simple. As with complex binary systems, the structural changes that take place upon temperature change in an alloy of given composition must be considered one at a time in the order in which they occur, and the influence of the structure as it exists just before each phase change must be taken into account in rationalizing the structural change that follows. Where a multiplicity of phase changes occurs in succession, without equilibrium conditions being approximated in any step, the resulting structures are likely to be very difficult to rationalize.

As a hypothetical example, consider the alloy of composition $X$ in the isothermal sections (of Fig. 18-2) that are presented in Fig. 18-5. If this alloy is heated above its melting point and then allowed to cool, freezing will begin with the primary separation of cored $\beta$ crystals (Fig. 18-6). At the temperature of the first section, $T_1$ in Fig. 18-5, primary crystallization is just beginning (Fig. 18-6a). Secondary crystallization of $\delta$ is proceeding at $T_3$ at the expense of both liquid and $\beta$, so that in Fig. 18-6b the $\beta$ has been partly eaten away and is encased in a thick layer of $\delta$. The $\delta$ phase disappears at temperature $T_4$ by decomposition into $\gamma$, $\beta$, and liquid (Fig. 18-6c). It is to be expected that this reaction, which is at relatively high temperature, will go to completion; the $\gamma$ phase should appear in places formerly occupied by $\delta$ but should share this space with the additional $\beta$ and liquid that is formed. Freezing of the remaining liquid is completed at $T_6$ (Fig. 18-6d) by isothermal reaction with $\beta$ to form $\gamma + \eta$, or if there is yet some liquid remaining because of incomplete reaction, it will freeze at lower temperature directly to $\gamma + \eta$. Finally at $T_9$, $\eta$, $\beta$, and $\gamma$ react to form $\omega$. All the $\eta$ phase would be consumed if the reaction were to go to completion. Peritectoid reaction at low tempera-

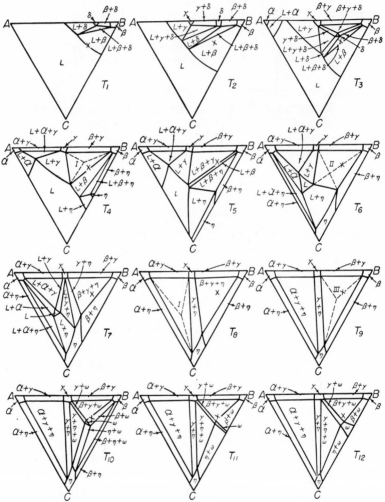

Fig. 18-5. Isotherms through the space diagram of Fig. 18-2.

ture is not likely, however, to go to completion. Consequently, it is presumed that *some* of the $\omega$ phase will form at the expense of each of the reacting phases but chiefly of the $\eta$ (Fig. 18-6e).

Assuming no further change in cooling to room temperature, the final cast structure of the alloy will be composed of a coarse primary $\beta$ constituent showing signs of peritectic attack along its edges, areas of the newly formed $\omega$ phase adjacent to the primary $\beta$ particles, and also smaller particles of a Widmanstätten-precipitated $\omega$ within the primary $\beta$ constituent, a residue of unreacted $\gamma$, and also some $\eta$ in the interdendritic spaces.

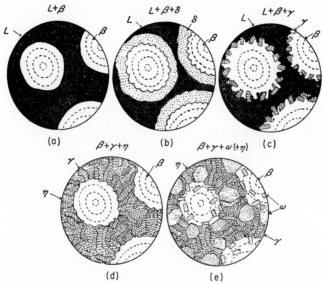

FIG. 18-6. Evolution of the microstructure of alloy $X$ of Fig. 18-5.

## Liquidus Projections

A convenient "shorthand" method for representing semiquantitatively the major features of a complex ternary system involves the use of simplified liquidus projections, such as that given in Fig. 18-7. Here only the traces of the valleys of the liquidus (bivariant equilibria) are plotted, and the directions of their slopes toward lower temperature are indicated by arrows inscribed upon the lines. If, in addition, the types of the univariant equilibria (intersections of the lines) are indicated by Roman numerals and the areas of tervariant equilibrium are labeled, a fairly comprehensive representation of the ternary system is obtained. Thus, in Fig. 18-7, point $a$ is clearly a eutectic point, $L \to \alpha + \beta + \gamma$; the Roman numeral I is really not essential here because the sloping of all three valleys toward point $a$ identifies it as a ternary eutectic. Points $b$ and $c$, being indicated as class II equilibria, evidently represent the reactions $L + \delta \to \alpha + \beta$ and $L + \eta \to \alpha + \delta$, respectively. In these cases, as at point $d$, two valleys approach from higher temperature and one proceeds to lower temperature. But point $d$ is indicated as class I equilibrium. Therefore, it must be one of the solid phases and not the liquid that undergoes eutectic-type decomposition; obviously the $\theta$ phase must be unstable below the temperature of $d$: $\theta \rightarrow \beta + \eta + L$. At point $e$ one valley approaches from higher temperature and two descend to lower temperature. This is a class III equilibrium $L + \beta + \eta \rightarrow \delta$ and represents the maximum temperature of existence of the $\delta$ phase.

Further insight into the meaning of this diagram and, at the same time, a check upon the self-consistency of the reactions shown may be had by deriving the plot shown on the opposite page and which is somewhat equivalent to that of Fig. 11-4.

The binary univariant equilibria (without reference to type) are read from the liquidus projection and written in the top line of the foregoing

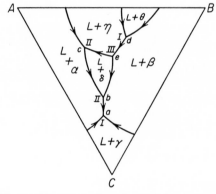

Fig. 18-7. Liquidus projection of a complex ternary system.

plot. In temperature sequence beneath, the ternary univariant equilibria are written and enclosed in "boxes." The four three-phase equilibria associated with each four-phase equilibrium are then written above and below each box, according to the type of four-phase equilibrium concerned. Thus, in the case of $L \rightarrow \alpha + \beta + \gamma$, the three three-phase equilibria involving the liquid phase are written above and the one all-solid three-phase equilibrium below. Next, matching pairs of three-phase groups involving the liquid phase are connected with lines and arrows pointing in the direction of lowering temperature. These represent the temperature spans of the three-phase fields and also the lines and arrows of Fig. 18-7. If the diagram is self-consistent, all three-phase groups involving the liquid phase will thus be paired; only groups of three solid phases will remain unpaired.

Since the projection of Fig. 18-7 represents only the boundaries of the liquid phase, this diagram tells nothing of the relationships obtaining in

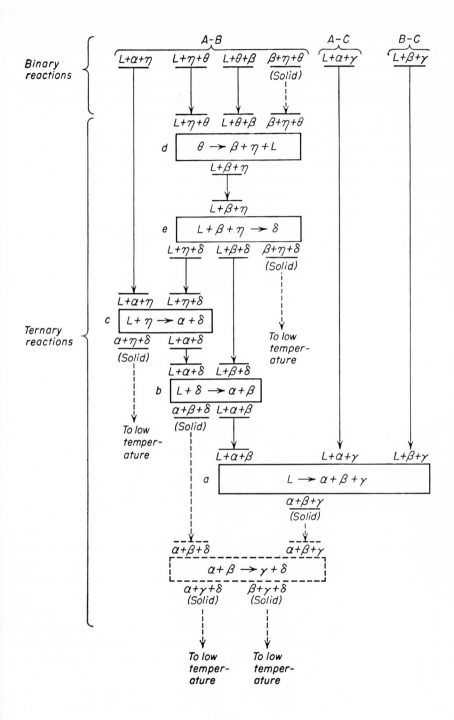

those portions of ternary space where all phases are solid. It provides, nevertheless, a basis for speculation concerning solid-state relationships and, taken in conjunction with the binary diagrams, limits the possibilities to some extent. To illustrate, consider the three-phase group $\beta + \eta + \theta$, involving the $\theta$ phase, which must disappear at the temperature of $d$. This group cannot be stable, therefore, below the temperature of point $d$, and this three-phase equilibrium must occur at higher temperature in the binary system $AB$. All the other solid-phase groups might be stable at room temperature. Assuming no further solid-state reactions, the room temperature isotherm should include the following three-phase fields: $\beta + \delta + \eta$, $\alpha + \delta + \eta$, $\alpha + \beta + \delta$, and $\alpha + \beta + \gamma$. All two-phase

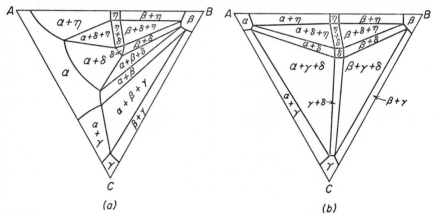

Fig. 18-8. Low-temperature isotherms of the system represented in Fig. 18-7(a) when no solid-state univariant reaction is assumed and (b) when the reaction $\alpha + \beta \rightarrow \gamma + \delta$ is assumed.

pairs that can be made from these three-phase groups will appear as two-phase equilibria, namely, $\beta + \delta$, $\beta + \eta$, $\delta + \eta$, $\alpha + \delta$, $\alpha + \eta$, $\alpha + \beta$, $\alpha + \gamma$, and $\beta + \gamma$. The one-phase fields at room temperature would then be $\alpha$, $\beta$, $\gamma$, $\delta$, and $\eta$. This would give a room temperature isotherm of the form shown in Fig. 18-8a.

This isotherm may be checked against the binary diagram of the system $AC$. Unless this binary diagram shows a very broad $\alpha$ field consistent with the isotherm that has been deduced, the evidence is strong that there must be one or more solid-state transformations. For example, the insertion of the transformation $\alpha + \beta \rightarrow \delta + \gamma$ (as indicated by the dashed construction in the foregoing chart) would relieve the necessity for a wide solid-solubility range in the binary system $AC$ (see Fig. 18-8b).

The information given by the liquidus projection can be augmented in several ways. Temperature contours may be shown, and the temperatures of univariant equilibria indicated. If, in addition, the compositions of the phases participating in the univariant equilibria are listed, the value of the diagram is considerably enhanced. This procedure is illustrated in Fig. 18-9, which is a liquidus projection of the complex diagram shown in Fig. 18-2. It will be seen that down to 600°C the projection gives almost as much information as does the space diagram itself.

Similar projections can, of course, be made of the boundaries of any single-phase field in the space diagram. Of particular usefulness are such projections as that of the $\gamma$ field in ternary diagrams of the iron-base

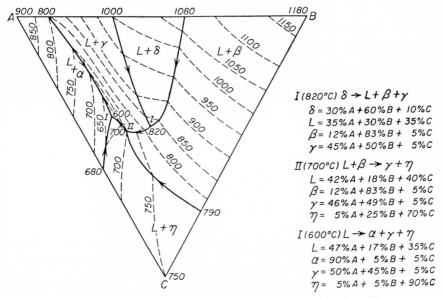

$I(820°C)\ \delta \rightarrow L + \beta + \gamma$
$\delta = 30\% A + 60\% B + 10\% C$
$L = 35\% A + 30\% B + 35\% C$
$\beta = 12\% A + 83\% B + 5\% C$
$\gamma = 45\% A + 50\% B + 5\% C$

$II(700°C)\ L + \beta \rightarrow \gamma + \eta$
$L = 42\% A + 18\% B + 40\% C$
$\beta = 12\% A + 83\% B + 5\% C$
$\gamma = 46\% A + 49\% B + 5\% C$
$\eta = 5\% A + 25\% B + 70\% C$

$I(600°C)\ L \rightarrow \alpha + \gamma + \eta$
$L = 47\% A + 17\% B + 35\% C$
$\alpha = 90\% A + 5\% B + 5\% C$
$\gamma = 50\% A + 45\% B + 5\% C$
$\eta = 5\% A + 5\% B + 90\% C$

FIG. 18-9. Liquidus projection of the space diagram of Fig. 18-2.

systems. By its use, the pattern of solid-state equilibria can be represented in a simple form. In some cases projections of the solidus have also been employed. These are less satisfactory, however, than liquidus projections and present little additional information.

In Chap. 14 a procedure was demonstrated for deducing the course of freezing of specific alloys by the use of the liquidus projection (Fig. 14-10). It will be recalled that the following principles are involved: (1) The liquid composition always changes toward lower temperature, (2) the liquid composition moves away from the compositions of the solid phases that are crystallizing, (3) but one solid phase is crystallizing as long as the liquid composition lies within one of the areas of the liquidus projection (i.e., two-phase region) and (4) two-phase crystallization occurs when the liquid composition is moving down one of the liquidus valleys.

## PRACTICE PROBLEMS

**1.** $A$ melts at 1000°C, $B$ at 800°C, and $C$ at 600°C; the system $AB$ has a peritectic reaction at 900°C; the system $AC$ has a eutectic reaction at 500°C; the system $BC$ has a eutectic reaction at 400°C; there is a ternary eutectic reaction at 300°C. Draw the space diagram, and develop isotherms at 950, 850, 550, 450, 300, and 250°C. Trace the course of freezing of an alloy containing 10% $A$ and 70% $B$.

**2.** In descending order of melting point the components of a ternary system are $A$, $B$, and $C$; each of the binary systems is of the simple eutectic type; there is a class II four-phase reaction in the ternary space of the diagram; no other isothermal reactions exist in this system. Draw the space diagram, and develop isothermal sections at each temperature level where a new configuration appears.

**3.** The ternary system $ABC$ of which the three binary diagrams are shown below has two class I four-phase reactions in the space diagram. Complete the space diagram, and develop isothermal sections to show the complete structure of the diagram.

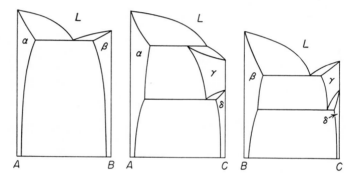

**4.** The ternary system $ABC$ of which the three binary diagrams are shown below has one class I and two class II four-phase reactions in the space diagram. Complete the space diagram, and develop isothermal sections to show the complete structure of the diagram.

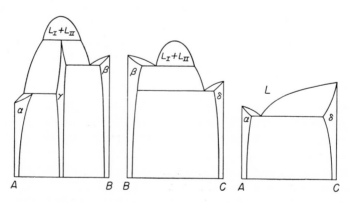

**5.** From the following vertical sections develop isothermal sections and the space diagram.

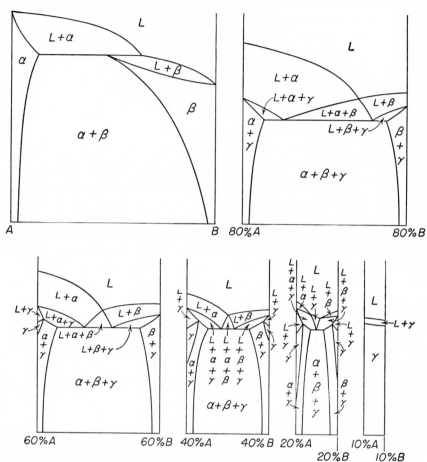

# CHAPTER 19

# MULTICOMPONENT SYSTEMS

From the standpoint of metallurgical practice, phase diagrams of alloy systems involving four and more elements are needed more than are the diagrams of simpler systems, because many of the alloys of commerce contain more than three elements added deliberately, and if the significant impurities are included in the count, the majority contain a still larger number of elements. It is unfortunate, therefore, that temperature-composition phase diagrams of four-component systems become extremely cumbersome and that those of more than four components are all complex to the point of being almost useless. About 40 quaternary systems (four components) have been investigated in part, none completely, and portions of perhaps a half dozen quinary systems (five components) have been studied. Hence, despite their complexity, it will be worth while to give the higher order systems some consideration, brief though it must necessarily be.

## Quaternary Systems

In order to express the composition of a quaternary alloy it is necessary to state the percentage or fraction of three of its four components. The graphical representation of these three variables together with the pressure and temperature (PTXYZ) would require a five-dimensional space. A temperature-concentration (TXYZ) diagram of a four-component system would require the use of four dimensions. Since neither is possible, it is necessary to resort to either three-dimensional isothermal isobaric sections or isobaric isopleths in which temperature and one or two concentration variables are represented in two or three dimensions, respectively. Both approaches will be discussed in this chapter.

For the representation of the three composition variables in three dimensions, an equilateral tetrahedron is employed (Fig. 19-1). This figure has the same properties in three dimensions as has the equilateral triangle in two dimensions. Four lines originating upon any one (composition) point within the tetrahedron, each drawn parallel to a different edge to intersection with one of the four faces of the figure, will have a total length equal to that of one edge of the tetrahedron. From point $P$, lying within the tetrahedron $ABCD$, Fig. 19-1, a line has been drawn parallel with $AD$ to its intersection with face $BCD$ at point $a$; another line $Pb$ has been drawn parallel to $AB$ and ends upon point $b$ in the face $ACD$; the line $Pc$ is drawn parallel to $AC$ and ends upon the face $ABD$; while $Pd$ is drawn to $ABC$ parallel to $BD$. The total length of these lines is equal to the length of any one of the edges such as $AB$. If, once more, the length of an edge be taken as 100%, then the composition at point $P$ is represented by the lengths of the four lines:

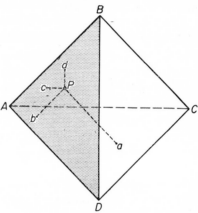

$$\%A = Pa$$
$$\%B = Pb$$
$$\%C = Pc$$
$$\%D = Pd$$

Pure components are represented at the corners of the tetrahedron, binary mixtures along the six edges, ternary mixtures upon the four faces, and quaternary mixtures within the space of the tetrahedron.

Were it mechanically feasible to do so, a three-dimensional grid for reading the composition of alloys, analogous to that employed with ternary diagrams in Fig. 12-2, might be constructed by passing regularly spaced planes through the tetrahedron parallel to each of its four faces.

FIG.1 9-1. Equilateral tetrahedron representing all possible compositions of the quaternary system $ABCD$. Point $P$ represents a quaternary composition within the tetrahedron, its position being shown by the four dashed lines which terminate upon the four faces of the tetrahedron.

This is not practical, however, with a two-dimensional drawing of a space model. Consequently, there is little that can be done, short of making actual space models, to make four-component isotherms quantitative. Several schemes for reading composition in two-dimensional drawings of space figures have been proposed, but none has been found satisfactory. The principal usefulness of the three-dimensional isotherm is, therefore, in exhibiting the structure of the quaternary diagram in a qualitative way. Isopleths may then be used to advantage to give the temperature and compositions at points distributed upon the lines and surfaces depicted in the isotherms. This is a cumbersome procedure, indeed, but nothing better is at hand.

## Representation of Equilibria in Quaternary Isotherms

According to the phase rule, there can be a maximum of five phases in equilibrium in a quaternary system after the pressure variable has been selected arbitrarily, thereby exercising one degree of freedom:

$$P + F = C + 2$$
$$5 + 1 = 4 + 2$$

Six distinct types of geometric construction are required to represent quaternary equilibria in isothermal sections. These are as follows:

*One-phase equilibrium* is represented in the isothermal tetrahedron by any space enclosed by two-phase boundary surfaces.

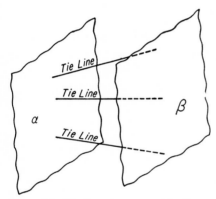

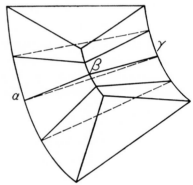

FIG. 19-2. Tie-lines in a space isotherm connect conjugate surfaces and may lie in any direction.

FIG. 19-3. Tie-triangles in a space isotherm connect three conjugate curves and may lie in any direction.

*Two-phase equilibrium* is represented in a space having two conjugate bounding surfaces connected at all points by tie-lines;[1] the tie-lines may run in any direction in the tetrahedral isotherm because every point in the figure is at the same pressure and temperature (see Fig. 19-2). Ruled boundaries of three-phase regions may partake in enclosing the two-phase space.

[1] This is true, of course, only of the isobaric isothermal section. In the (imaginary) TXYZ diagram tie-lines would connect two conjugate three-dimensional surfaces (i.e., conjugate volumes), tie-triangles would connect three conjugate *surfaces*, and tie-tetrahedra would connect four conjugate lines. In the complete PTXYZ diagram tie-lines would connect pairs of conjugate four-dimensional volumes, tie-triangles would connect sets of three three-dimensional volumes, tie-tetrahedra would connect sets of four conjugate surfaces, and the five-phase tetrahedra and hexahedra would connect sets of five conjugate curves. In drawing isopleths and projections these differences should be remembered.

*Three-phase equilibrium* is represented by three conjugate curves everywhere connected by tie-triangles;[1] these, too, may be oriented in any manner (see Fig. 19-3). Bounding surfaces of three-phase regions are shared with two-phase and four-phase regions.

*Four-phase equilibrium* is represented by a tie-tetrahedron[1] the four corners of which touch the four one-phase regions involved in the equilibrium. The tie-tetrahedron can have any shape that can be constructed by the use of four triangular faces of an assortment of dimensional ratios, and it may be oriented in any way (see Fig. 19-4). It should be noted that any ternary four-phase reaction plane, whether class I, II, or III, is the limiting (two-dimensional) case of a tetrahedron (three-dimensional).

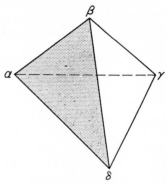

Fig. 19-4. Tie-tetrahedron associating the four phases $\alpha$, $\beta$, $\gamma$ and $\delta$.

Figure 19-5 represents the development of a tetrahedron from the class I or III reaction triangle, and Fig. 19-6 illustrates the development of a tetrahedron from the reaction trapezium of class II.

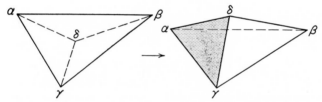

Fig. 19-5. Development of a tie-tetrahedron from the ternary class I, or class III, four-phase reaction triangle.

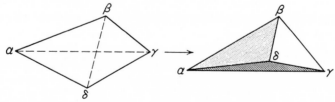

Fig. 19-6. Development of a tie-tetrahedron from the ternary class II four-phase reaction trapezium.

*Five-phase equilibrium* is represented by a tie-tetrahedron[1] (Fig. 19-7a) or a tie-hexahedron[1] (Fig. 19-7b), that is isothermal and, hence, can occur in only one of a series of isotherms taken at different temperatures. There are four classes of quaternary five-phase equilibrium, two of which

[1] See footnote, on opposite page.

are represented in Figs. 19-8 and 19-9 and the other two of which are the geometric inverse of the first two.

*Class I five-phase equilibrium* is represented by four four-phase tie-tetrahedra descending from higher temperature and joining to form the five-phase isothermal tetrahedron from which a single four-phase tie-tetrahedron descends to lower temperature. The fifth phase is represented at that point within the five-phase tetrahedron where the four

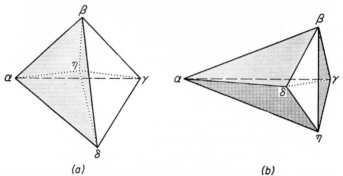

(a)                                        (b)

Fig. 19-7. Two possible kinds of space figures representing five-phase equilibrium: (a) tetrahedron, (b) hexahedron.

apexes of the four four-phase tetrahedra meet, as at $\eta$ in Fig. 19-8. The quaternary eutectic is a reaction of the first class:

$$L \underset{\text{heating}}{\overset{\text{cooling}}{\rightleftarrows}} \alpha + \beta + \gamma + \delta$$

as also is the quaternary eutectoid:

$$\alpha \underset{\text{heating}}{\overset{\text{cooling}}{\rightleftarrows}} \beta + \gamma + \delta + \eta$$

There are in all 23 combinations of phases that may conceivably partake in this class of reaction.

*Class II five-phase equilibrium* is represented by three four-phase tie-tetrahedra descending from higher temperature and joining to form an isothermal tie-hexahedron at the five corners of which are represented the five phases that coexist in isothermal equilibrium. This figure divides into two tie-tetrahedra that descend to lower temperature (Fig. 19-9). The reaction is somewhat analogous to ternary class II four-phase reaction:

$$L + \alpha \underset{\text{heating}}{\overset{\text{cooling}}{\rightleftarrows}} \beta + \gamma + \delta$$

There are 29 combinations of phases that could conceivably partake in class II five-phase equilibrium.

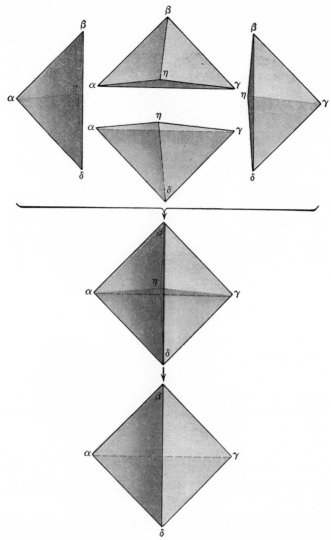

Fig. 19-8. Schematic representation of class I five-phase equilibrium. The geometric inverse of this is class IV equilibrium.

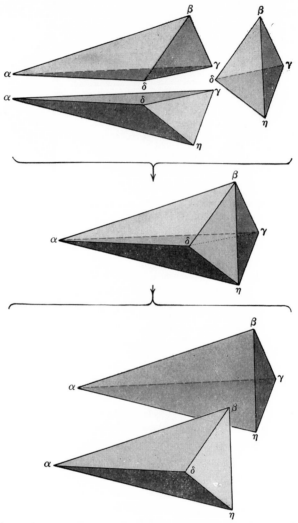

FIG. 19-9. Schematic representation of class II five-phase equilibrium. The geometric inverse of this is class III equilibrium.

*Class III five-phase equilibrium* is the inverse of class II equilibrium. Two tie-tetrahedra descending from higher temperature join to form an isothermal tie-hexahedron which then divides into three tie-tetrahedra that continue to lower temperature:

$$L + \alpha + \beta \underset{\text{heating}}{\overset{\text{cooling}}{\rightleftharpoons}} \gamma + \delta$$

*Class IV five-phase equilibrium* is, similarly, the inverse of class I equilibrium. A single tie-tetrahedron descending from higher temperature becomes the five-phase tie-tetrahedron. Reaction to form a new phase divides the tetrahedron into four tetrahedral parts that proceed to lower temperature. This is the quaternary peritectic reaction:

$$L + \alpha + \beta + \gamma \underset{\text{heating}}{\overset{\text{cooling}}{\rightleftharpoons}} \delta$$

and the same reaction involving only solid phases is the quaternary peritectoid:

$$\alpha + \beta + \gamma + \delta \underset{\text{heating}}{\overset{\text{cooling}}{\rightleftharpoons}} \eta$$

## Temperature-composition Sections

Isopleths in which a limited series of alloy compositions is represented, with the temperature variable, may be taken in a number of ways. For example, the composition with respect to one of the four components may be fixed, while the compositions of the other three are permitted to vary. Compositions designated by fixing the $B$ content at 40% are contained in the plane triangle $XYZ$ of Fig. 19-10. If this plane is made the base of a triangular prism, as in Fig. 19-17, coordinates resembling those employed in the construction of ternary space models are obtained. The diagrams so obtained often resemble ternary diagrams in a superficial way, but it should be remembered that they include only a small fraction of the alloys of the quaternary system and that tie-lines, tie-triangles, and tie-tetrahedra will not, in general, lie wholly within such a section but will extend through a series of such sections.

Another type of composition section in which the ratio of two of the components is held constant is shown in Fig. 19-11, where the plane $ADX$ corresponds to a constant ratio of $B$ and $C$. Such sections are useful chiefly in representing "quasi-ternary" equilibria, when $X$ is a congruently melting intermediate phase that forms a simple ternary system with $A$ and $D$. If a constant ratio is maintained between one of the components and two of the other components, the section appears as in plane $AXY$ of Fig. 19-12. A quasi-ternary system between two intermediate phases and a component may be represented on this type of section.

Diagrams of sections that can be drawn in two dimensions are obtained by limiting the composition series to a single line through the quaternary composition tetrahedron. It is usual to take the lines parallel to one of the edges of the space figure as line $XY$ in Fig. 19-13. This line represents all

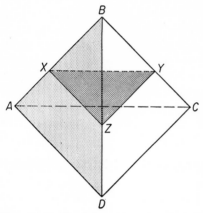

FIG. 19-10. Shaded section represents a system of alloys of a fixed $B$ content.

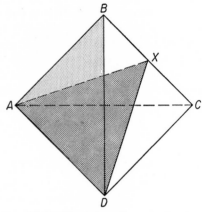

FIG. 19-11. Shaded section represents a system of alloys having a fixed ratio of the $B$ and $C$ components.

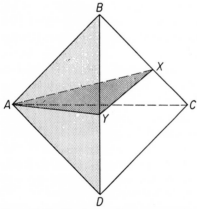

FIG. 19-12. Shaded section represents a system of alloys in which there is a fixed ratio between $B$ and the sum of the $C$ and $D$ components.

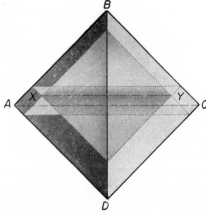

FIG. 19-13. Line $XY$, lying in a plane of constant $B$ content, represents a constant $D$ content in that plane; i.e., only the $A$ and $C$ contents vary along the line.

compositions containing 20% $B$ and 10% $D$; the percentages of $A$ and $C$ then vary between 0 and 70. If this line is made the base line of a set of coordinates similar to those employed in the representation of binary equilibria, diagrams of the kind shown in Fig. 19-18 are obtained. Once again it must be emphasized that the several tie-elements are not generally included in such sections and that the compositions of the participating phases are indicated only in certain special cases.

## Quaternary Isomorphous Systems

The best qualitative view of the equilibria in a quaternary isomorphous system is obtained from isotherms as in Fig. 19-14. The six tetrahedra in this illustration are arranged in order of descending temperature from $T_1$ to $T_6$. Component $A$ melts above $T_1$, component $C$ between $T_3$ and $T_4$,

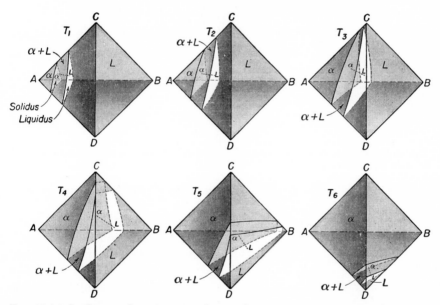

FIG. 19-14. Isotherms of a quaternary isomorphous system, temperature decreasing from $T_1$ to $T_6$.

component $B$ between $T_5$ and $T_6$, and component $D$ below $T_6$. Each isotherm shows a section of liquidus and a section of solidus. The liquidus and solidus surfaces sweep downward in temperature in successive isotherms from the melting point of $A$ to that of $D$. At each component (corner of the tetrahedron) the liquidus and solidus meet. Tie-lines connect the conjugate liquid and solid phases at every point on this pair of surfaces. The short dotted lines in each isotherm, labeled "$\alpha L$," are typical tie-lines included to show how these lines turn as the temperature is lowered. It should be noted that they are not necessarily horizontal as are the tie-lines in ternary space models, because the entire tetrahedron is isothermal. Tie-lines near the edges of the tetrahedron are nearly parallel with the edges, and the transition in direction is more or less

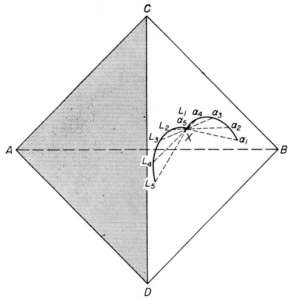

FIG. 19-15. Projection of the tie-lines involved in the freezing of a quaternary solid solution alloy of composition $X$.

regular in crossing the diagram. The usual lever principle is applicable to tie-lines in quaternary systems.

Freezing follows a curved path as shown in Fig. 19-15. This drawing is a

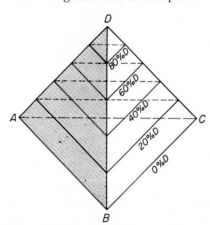

FIG. 19-16. Plan of composition sections represented in the space isopleths of Fig. 19-17.

liquidus and solidus "projection" analogous to that shown for the ternary case in Fig. 12-13. The gross composition of the alloy under consideration is indicated at $X$. Passing through point $X$ is a series of tie-lines covering the temperature range of freezing. Freezing begins at $T_1$, with liquid and solid represented at the ends of the tie-line $\alpha_1 X$, and is completed at $T_5$, the tie-line $X L_5$. Coring is possible as in simpler systems, the microstructure of a cored quaternary solid solution being indistinguishable from that of binary and ternary alloys.

A series of isopleths representing constant percentages of the $D$ component, Fig. 19-16, is presented in Fig. 19-17. In order to show the relationship between these and the isothermal sections of Fig. 19-14, each

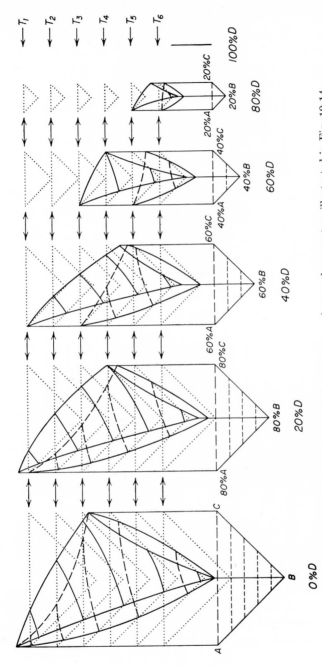

$T_1$
$T_2$
$T_3$
$T_4$
$T_5$
$T_6$

20%C
100%D
20%B
80%D
20%A

40%C
40%B
60%D
20%A

60%C
60%B
40%D
40%A

80%C
80%B
20%D
60%A

C
80%A

B

A

0%D

Fig. 19-17. Space isopleths through the quaternary isomorphous system illustrated in Fig. 19-14.

231

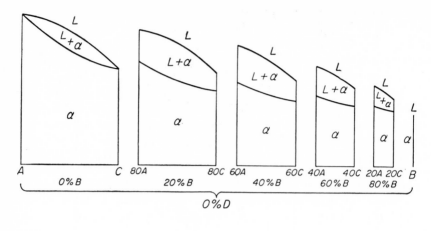

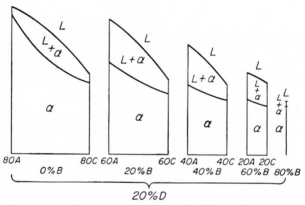

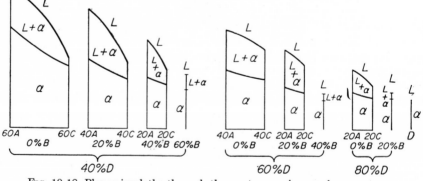

Fig. 19-18. Planar isopleths through the quaternary isomorphous system.

model has been intersected by horizontal dotted planes at the temperatures $T_1$ to $T_6$. The intersection of each of these planes with the liquidus and solidus surfaces corresponds in temperature and composition with an intersection of a plane, such as 20% $D$ in Fig. 19-16, with the liquidus and solidus in one of the isotherms of Fig. 19-14. The isopleth on the left of Fig. 19-17 is, in fact, the ternary system $ABC$, and the liquidus and solidus meet at the three corners of the prism. All others are quaternary sections so that there is no meeting of liquidus and solidus at corners. At 100% $D$ the section shrinks to a single line the height of which indicates the melting point of $D$.

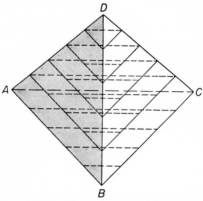

With both the $D$ and $B$ contents held constant, the two-dimensional isopleths of Fig. 19-18 are obtained. The sections represented are indicated by the dashed lines drawn upon the base of each prism in Fig. 19-17 and also by the dashed lines in Fig. 19-19. Six series of these diagrams, representing 0, 20, 40, 60, 80, and 100% $D$, are presented in Fig. 19-18. Individual diagrams in each series represent constant $B$ content at intervals of 20%. Liquidus and solidus temperatures corresponding

FIG. 19-19. Plan of composition series represented in the isopleths of Fig. 19-18.

to definite gross composition points may be measured on these diagrams, but the conjugate liquid and solid compositions cannot be associated.

Twenty-one separate sketches have here been used to depict the quaternary diagram at composition intervals of 20%. This is obviously an insufficiently complete survey for practical purposes. In order to decrease the composition intervals to 5%, which would be fairly satisfactory for practical uses, it would require the determination of 231 sections! There is little wonder that no fully determined quaternary diagrams are available.

## An Example Involving Three-phase Equilibrium

An example of the representation of three-phase equilibrium without further complications may be found in a quaternary system in which three of the binary systems, $AB$, $BC$, and $AC$, are isomorphous and three, $AD$, $BD$, and $CD$, are of the simple eutectic type (Fig. 19-20). The first isotherm $T_1$, Fig. 19-20$a$, is taken at the $BD$ binary eutectic temperature; the liquidus surfaces emerging from the $B$ and $D$ corners have just met on the line $BD$. At a slightly lower temperature $T_2$ tie-triangles $L + \beta + \delta$

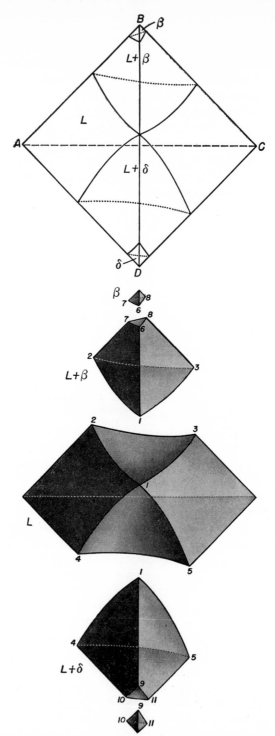

FIG. 19-20a. Six isotherms of a quaternary system in which the three-phase equilibrium $L + \beta + \delta$ sweeps through the diagram, with falling temperature, from the binary eutectic $BD$ at $T_1$ to the binary eutectic $AD$, which lies between $T_5$ and $T_6$.

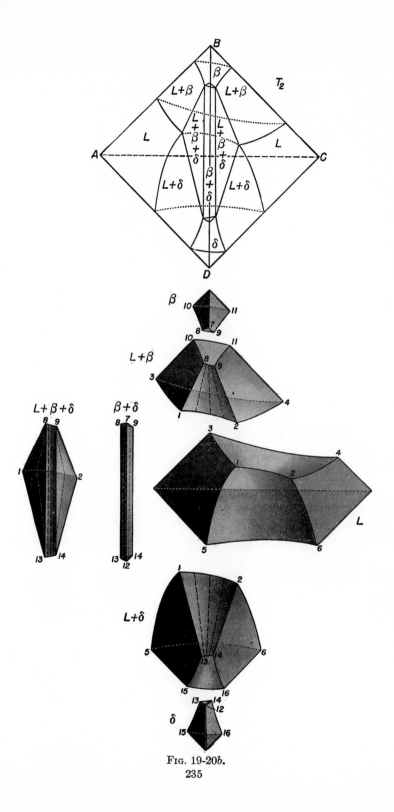

FIG. 19-20b.

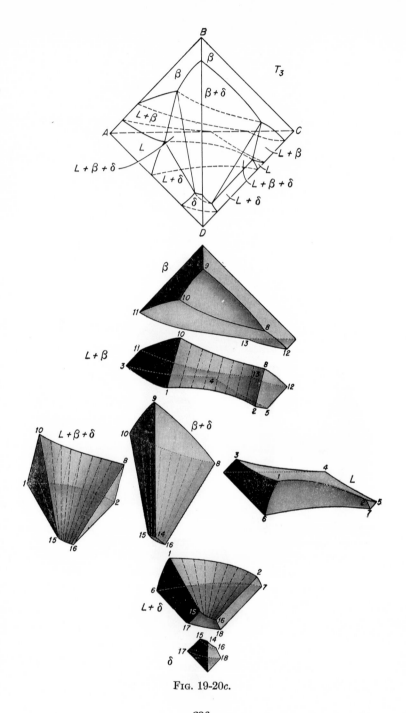

FIG. 19-20c.

236

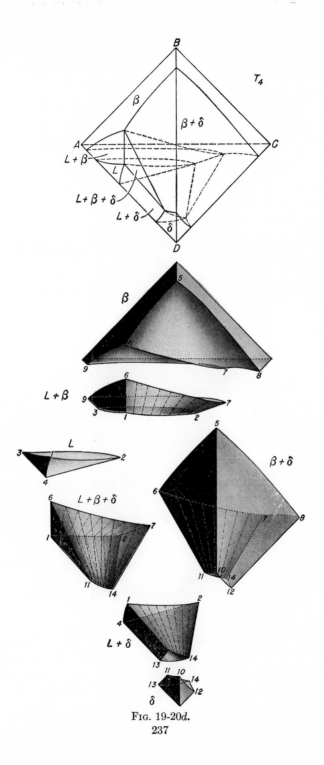

FIG. 19-20d.

237

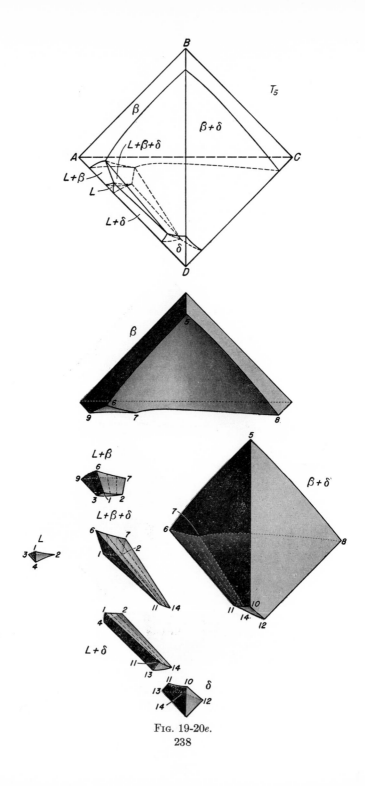

FIG. 19-20e.

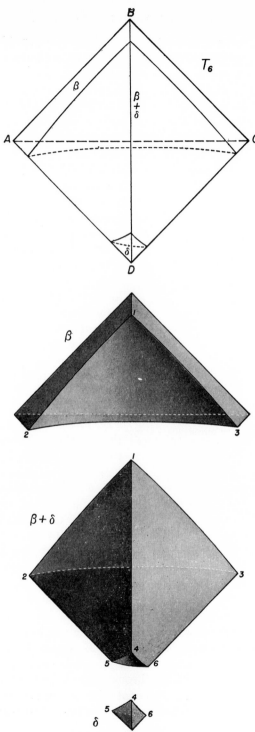

$T_6$

$\beta$

$\beta + \delta$

$\beta + \delta$

$\delta$

Fig. 19-20f.

239

are seen in the ternary faces $ABD$ and $BCD$ (see Fig. 19-20$b$). Indeed, these two faces appear very much the same as isotherm $T_2$ in Fig. 13-6. The three corners of the two tie-triangles in the ternary faces are connected point for point to form a solid space of triangular cross section (the space field $L + \beta + \delta$); i.e., the two liquid points are connected as are the two $\beta$ points and the two $\delta$ points. This region is shown at the extreme left in the exploded model. An elongated region of $\beta + \delta$ lies between the $BD$ edge of the diagram and the $L + \beta + \delta$ region, forming one surface of that region. Another surface of the three-phase region is formed by the top boundary of the $L + \delta$ region, and the third surface is the front boundary of the $L + \beta$ region. All these surfaces are inscribed with tie-lines in the exploded model; also, to aid the reader, the corners of the several regions are identified by numbers that correspond from region to region.

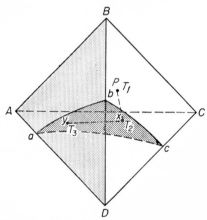

FIG. 19-21. Liquidus projection of the quaternary diagrams of Fig. 19-20. The shaded plane is the trace of all compositions of the liquid phase in three-phase equilibrium with $\beta$ and $\delta$, i.e., the quaternary equivalent of the ternary "liquidus valley." The dashed line represents the course of composition change of the liquid phase during the freezing of the alloy of initial composition $P$.

As the temperature continues to fall, the three-phase region becomes broader and approaches the lower face of the tetrahedron (see Fig. 19-20$c$). Between $T_3$ and $T_4$ it meets the binary eutectic of the system $CD$, where the terminal cross section of the region is reduced to a single line. From $T_4$, Fig. 19-20$d$, downward in temperature, the $L + \beta + \delta$ region no longer touches the $BCD$ face of the tetrahedron, all alloys of this ternary system now being solid. It extends, instead, from the $ACD$ face to the $ABD$ face. Finally, between $T_5$, Fig. 19-20$e$, and $T_6$, Fig. 19-20$f$, the three-phase region ends in the $L + \beta + \delta$ binary eutectic line of the system $AD$, and the last of the liquid disappears. Below the $AD$ eutectic the quaternary system is seen to be composed of only three regions, namely, $\beta$, $\delta$, and $\beta + \delta$.

The course of freezing of a typical alloy may be followed with these isotherms, but a simpler view is obtained by the use of a "liquidus projection." The shaded surface $abc$ in Fig. 19-21 represents the trace of the "liquidus valley" (i.e., trace of the line representing liquid on the $L + \beta + \delta$ region in Fig. 19-20) through the entire freezing range. Consider the behavior of a four-component alloy of composition $P$, which lies within the space of the tetrahedron between the $BC$ edge and the surface *abc*. As freezing begins at $T_1$, the liquid will have the composition $P$. Primary crystals of $\beta$ are rejected, and the liquid composition moves downward and forward, away from the $B$ corner, with falling temperature, until at $T_2$ it reaches point $x$ on the surface *abc*. Now a secondary separation of $\beta + \delta$ begins. The direction of shift of the liquid composition changes sharply and moves away from the $BD$ edge along the surface *abc* to $y$ at $T_3$, where the liquid is exhausted and the alloy is entirely solid. This alloy will have a microstructure which appears very similar to that of a binary hypoeutectic alloy (Fig. 4-8*b*) or of a ternary alloy involving only three-phase equilibrium

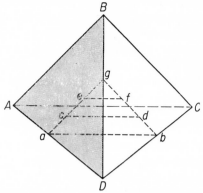

Fig. 19-22. Plan of isopleths shown in Fig. 19-23.

(Fig. 13-9, 20°C). The coring will, of course, be more complex than appears in the microstructure, since there is variation with respect to four

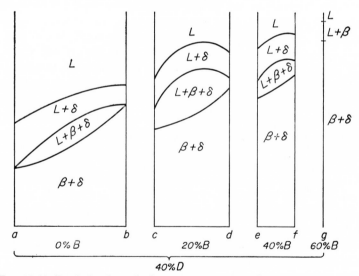

Fig. 19-23. Isopleths through the quaternary phase diagram of Fig. 19-20.

components. An analysis of the tie-triangles in this diagram will show that both of the solid phases must increase in quantity as the liquid diminishes during cooling. With other reactions combined in three-phase equilibrium, corresponding transformation characteristics can be developed as was done in the case of ternary alloys in Chap. 13.

In order to record actual temperatures of phase changes in this system, as with all quaternary alloys, it is necessary to employ a series of isopleths. A typical series taken at 40% $D$ with fixed percentages of $B$ in successive diagrams is given in Fig. 19-23. A key to the location of these sections appears in Fig. 19-22. The first diagram of the series $a$-$b$ lies in the ternary system $ACD$ and is normal for a vertical section through a ternary alloy. Succeeding sections are somewhat similar, though it should be noted that the three-phase field is not closed at the edges of the diagram.

## A System Involving Four-phase Equilibrium

A simple example of four-phase equilibrium in a quaternary system is illustrated in the seven isotherms of Fig. 19-24. Among the six binary systems concerned, five are of the simple eutectic type and one $(AC)$ is isomorphous; two of the four ternary systems are of the simple ternary eutectic type, while the other two involve no four-phase equilibria.

The first isotherm in Fig. 19-24$a$ is taken at a temperature slightly below the $BD$ binary eutectic at which the three-phase region $L + \beta + \delta$ originates. Just above the $ABD$ ternary eutectic temperature, $T_2$ in Fig. 19-24$b$, three three-phase regions are seen growing toward a junction near the center of the $ABD$ face of the tetrahedron. The junction is effected with the first appearance of four-phase equilibrium, $L + \alpha + \beta + \delta$, at the ternary eutectic temperature $T_3$, Fig. 19-24$c$. Point $L$ of the ternary four-phase equilibrium now moves rapidly into the space diagram, while points $\alpha$, $\beta$, and $\delta$ move slowly inward, Fig. 19-24$d$. Thus is created a tetrahedral figure representing quaternary four-phase equilibrium. Upon each of its four faces it meets a three-phase region, $\alpha + \beta + \delta$ originating upon the $ABD$ face of the isotherm, $L + \alpha + \beta$ upon the $ABC$ face, $L + \alpha + \delta$ upon the $ACD$ face, and $L + \beta + \delta$ upon the $BCD$ face.

As the temperature continues to fall, this four-phase tetrahedron sweeps through the space diagram (see $T_5$, Fig. 19-24$e$) approaching the $BCD$ ternary eutectic $(T_6$, Fig. 19-24$f)$. The last of the liquid vanishes as the four-phase tetrahedron closes to a plane and terminates upon the ternary eutectic isotherm at the $BCD$ ternary eutectic temperature. At this and lower temperatures a three-phase region $\alpha + \beta + \delta$ crosses the space diagram from its $ABD$ face to its $BCD$ face (Fig. 19-24$g$).

Alloys of this system should be expected to develop structures similar to ordinary ternary eutectic alloys except that coring is possible in all constituents. A series of typical isopleths taken at 20% of the $B$ component and with the quantity of the $D$ component increasing in intervals of 20% in the series is given in Fig. 19-25. It will be noted that there are no isothermal (horizontal) boundaries anywhere in these diagrams.

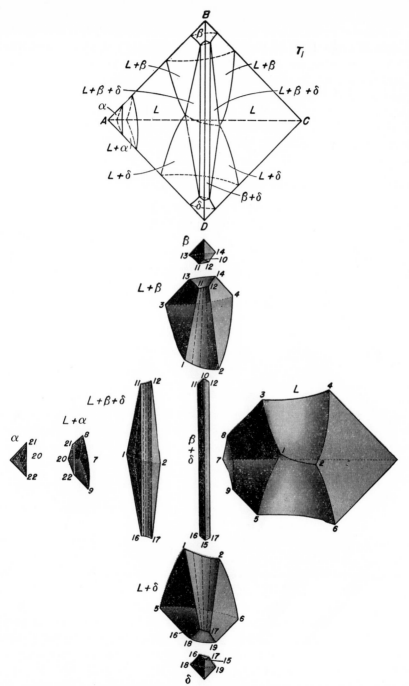

FIG. 19-24a. Seven isotherms from a quaternary system in which the four-phase equilibrium $L + \alpha + \beta + \delta$ sweeps through the diagram from the ternary eutectic $ABD$ at $T_3$ to the ternary eutectic $BCD$ between $T_6$ and $T_7$. The sequence of isotherms is from high temperature at $T_1$ to low temperature at $T_7$.

243

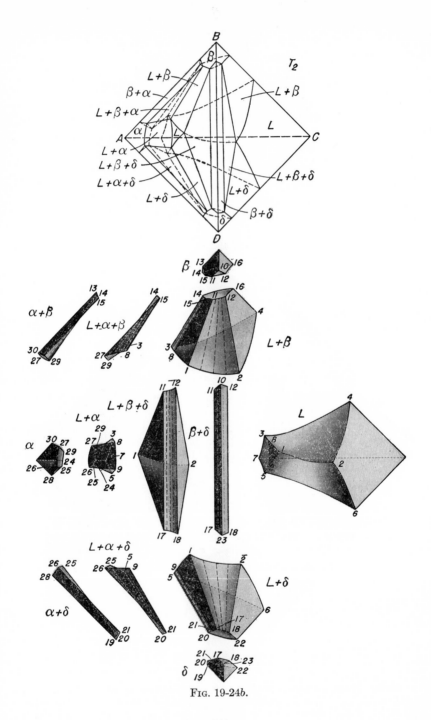

FIG. 19-24b.

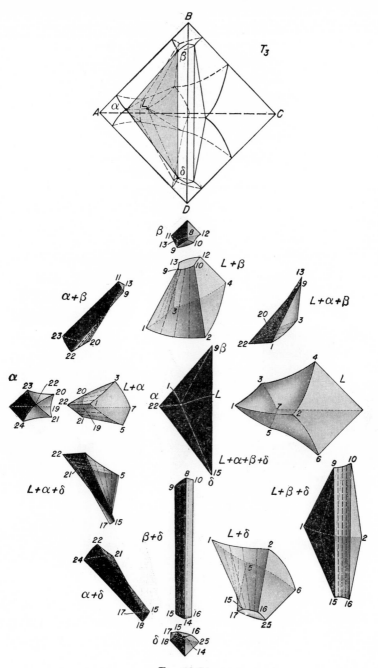

FIG. 19-24c.

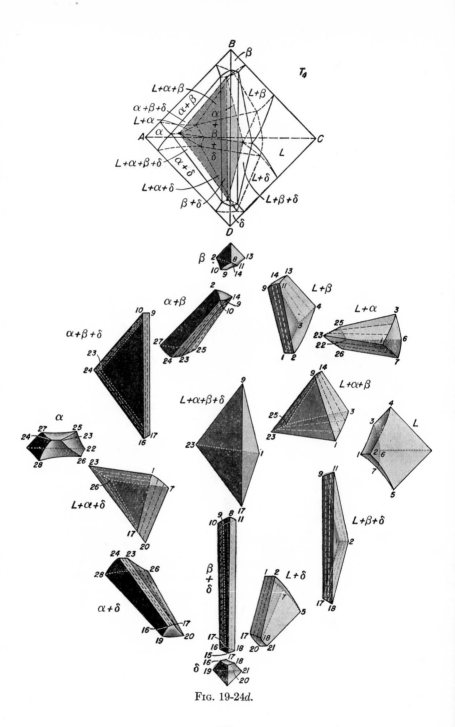

F<small>IG</small>. 19-24*d*.

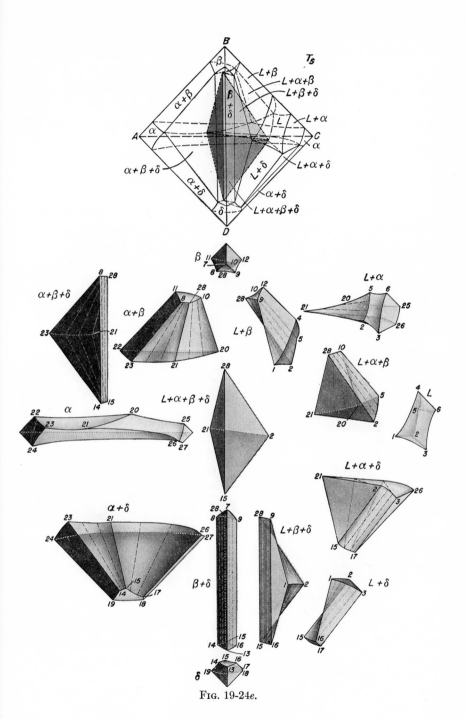

Fig. 19-24e.

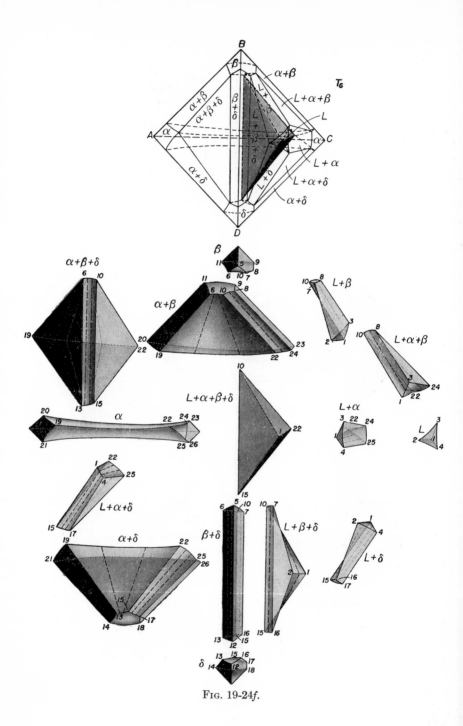

FIG. 19-24f.

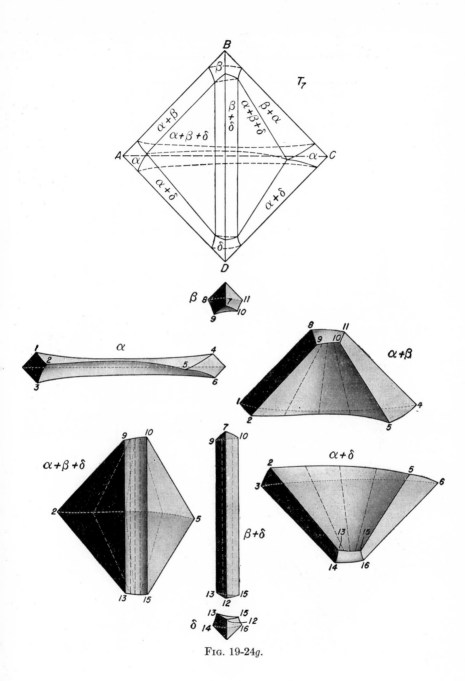

$T_7$

$\beta$

$\alpha$

$\alpha + \beta$

$\alpha + \beta + \delta$

$\beta + \delta$

$\delta$

FIG. 19-24g.

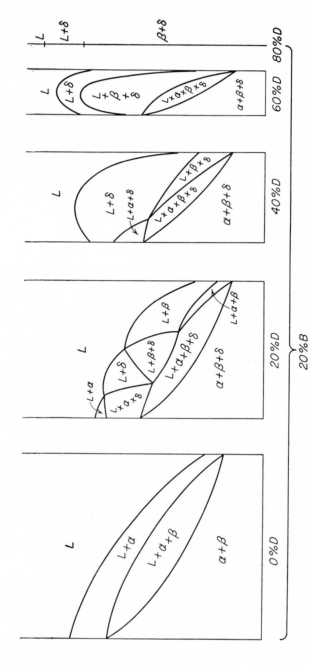

FIG. 19-25. Isopleths through the quaternary system shown in Fig. 19-24.

## An Example of Class I Five-phase Equilibrium, A Quaternary Eutectic System

Five-phase equilibrium of the first kind may be exemplified by a *quaternary eutectic* system, Fig. 19-26. This rather complex diagram is, perhaps, best understood by comparing it with the foregoing example of four-phase equilibrium. Here, all the six binary systems are taken to be of the eutectic type, and all four ternary systems are of the simple ternary eutectic class. The first six isotherms of Fig. 19-26 record the progress of freezing the four ternary eutectics with the development of a four-phase tetrahedron corresponding to each. At $T_2$, Fig. 19-26*b*, the ternary eutectic of the system $ABD$ is being approached. This reaction results in the formation of the tetrahedral region $L + \alpha + \beta + \delta$ that appears for the first time in section $T_3$, Fig. 19-26*c*. Between $T_3$ and $T_4$ the $ACD$ ternary eutectic is passed, introducing the tetrahedral region $L + \alpha + \gamma + \delta$, Fig. 19-26*d*, and between $T_4$ and $T_5$ the ternary eutectic of the system $ABC$ gives rise to the four-phase region $L + \alpha + \beta + \gamma$, Fig. 19-26*e*. All four-phase regions are present in section $T_6$, Fig. 19-26*f*, which is below all the ternary eutectic temperatures but above the quaternary eutectic temperature.

The liquid region at $T_6$ is enclosed wholly within the space of the isotherm, being outlined by six curved lines that join the liquid points of the four four-phase tie-tetrahedra. This region shrinks to a single point at the quaternary eutectic temperature $T_7$, Fig. 19-26*g*, where the four four-phase tetrahedra join to form the five-phase isothermal tetrahedron that is the *quaternary eutectic reaction isotherm*. Below $T_7$ the configuration of the diagram remains qualitatively constant, with a single tie-tetrahedron representing the equilibrium $\alpha + \beta + \gamma + \delta$.

The *quaternary eutectic point* is the point in composition that represents the last liquid to freeze. An alloy of this composition freezes isothermally with the simultaneous rejection of four solid phases. The resulting structure is illustrated for the quaternary eutectic alloy of the system bismuth-cadmium-lead-tin (approximately "Wood's metal") in Fig. 19-27. In this case, the bismuth, being present in major proportion, forms a matrix in which the other three phases are embedded. It is worthy of note that equilibrium is achieved in complex structures such as this without each phase necessarily making physical contact with every other phase. If $\alpha$ is in contact with $\beta$ and the two are in mutual equilibrium, and if, at the same time, $\gamma$ is in contact and in equilibrium with $\beta$, then $\alpha$ and $\gamma$ will be in equilibrium with each other whether or not they touch each other in the microstructure.

Alloys of other than quaternary eutectic composition may freeze in a variety of ways. The pattern of freezing of a randomly selected quaternary

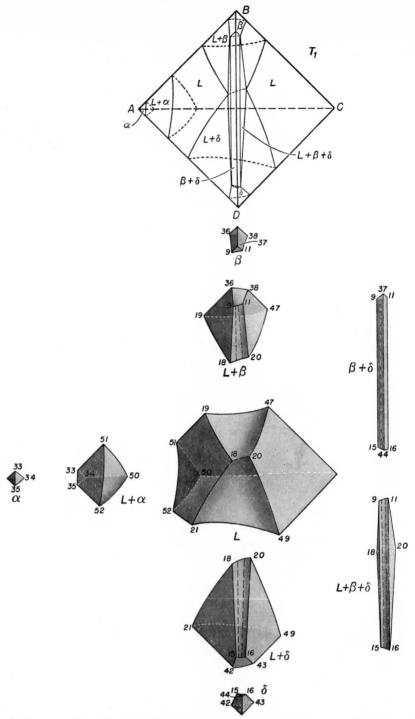

FIG. 19-26a. Eight isotherms showing, in descending temperature, the sequence of changes in a quaternary eutectic system. The quaternary eutectic itself lies at $T_{1}$ where the five-phase equilibrium $L + \alpha + \beta + \gamma + \delta$ is represented.

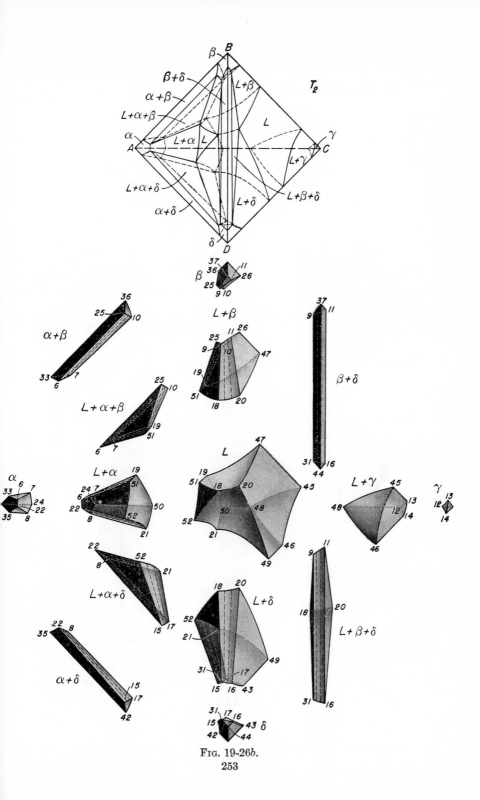

Fig. 19-26*b*.
253

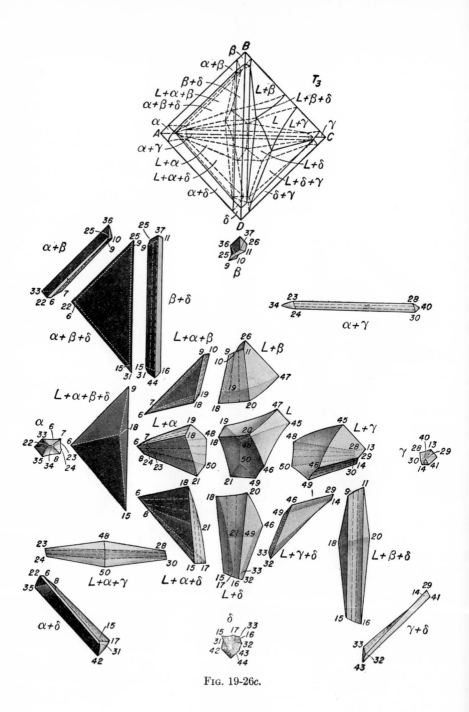

FIG. 19-26c.

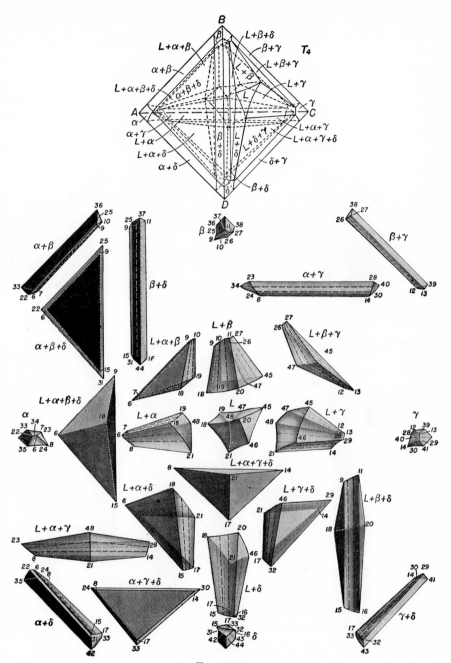

FIG. 19-26d.

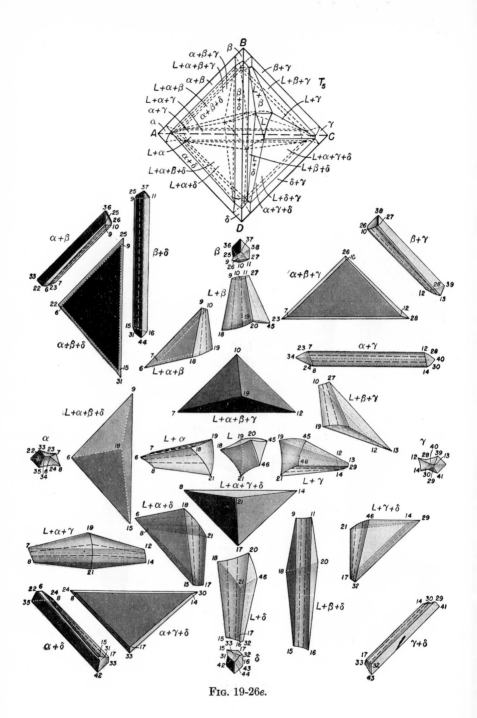

FIG. 19-26e.

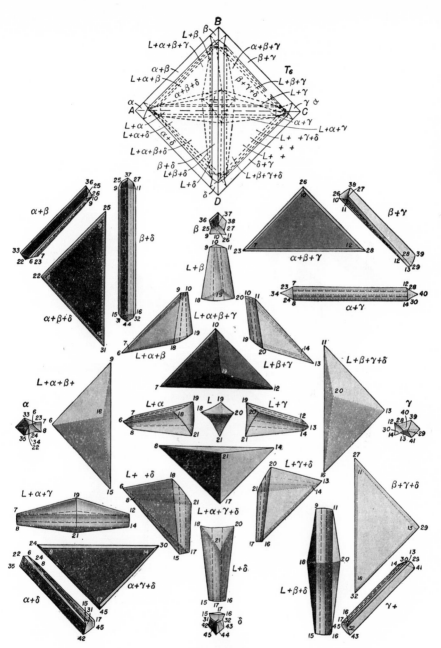

FIG. 19-26*f*.

257

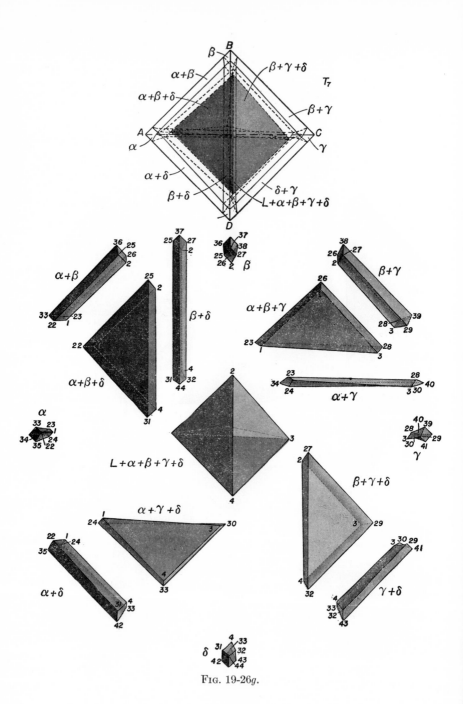

FIG. 19-26g.

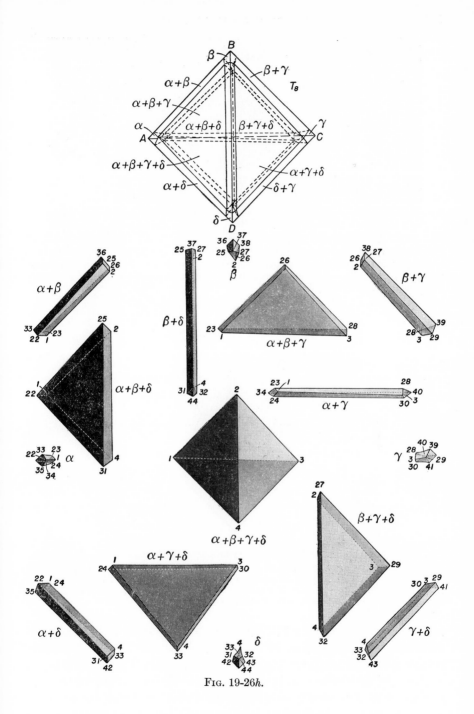

Fig. 19-26h.

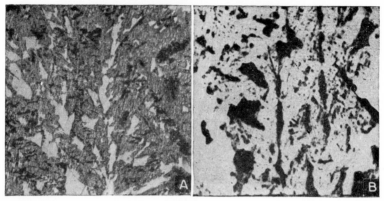

Fig. 19-27. Microstructure of the quaternary eutectic of the system bismuth-lead-tin-cadmium; the four phases are, respectively: bismuth (light-colored matrix, best seen in B); lead (small, dark, script-like particles, giving gray color to the matrix in A); tin (large, white masses, most evident in A); and cadmium (large, black particles). Magnification: A, 100; B, 500.

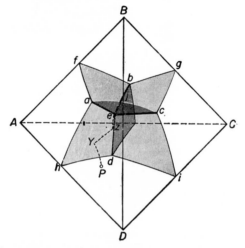

Fig. 19-28. Liquidus projection of the quaternary eutectic system. The dashed line, originating upon the composition of alloy $P$, traces the change in the composition of the liquid phase during the primary crystallization of $\delta$ ($P$ to $Y$), the secondary crystallization of $\alpha + \delta$ ($Y$ to $Z$), the tertiary crystallization of $\alpha + \gamma + \delta$ ($Z$ to $e$), and the final freezing of the last liquid of eutectic composition $e$.

composition, such as $P$ in Fig. 19-28, may be followed qualitatively by reference to a "projection of the liquidus valleys." Lines of ternary bivariant equilibrium involving the liquid phase (the liquidus valleys, for example, $fa$) are shown as light lines in the ternary faces of the tetrahedron of Fig. 19-28. Quaternary bivariant lines involving liquid, for example, $ae$, in the space of the diagram (the traces of the "liquid points" of the

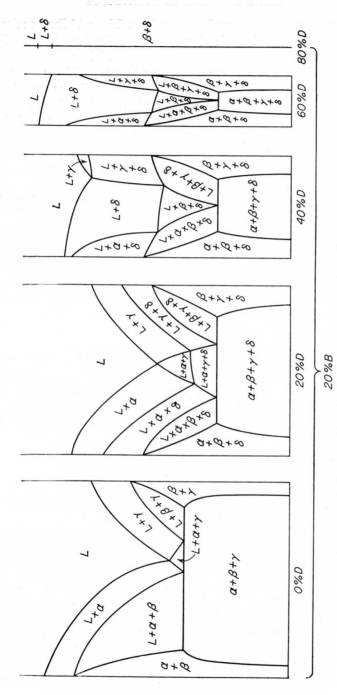

Fig. 19-29. Isopleths through the quaternary eutectic diagram of Fig. 19-26.

261

four tie-tetrahedra) are shown in heavy lines. These 16 lines enclose six tervariant surfaces, such as *haed*, representing the traces of the liquid points on three-phase equilibria involving liquid. The alloy of composition $P$ first rejects primary $\delta$, and the liquid composition moves away from the $D$ corner of the diagram until it meets the surface *haed* at point $Y$. Here, $\alpha$ and $\delta$ freeze cooperatively as a two-phase secondary constituent and the liquid composition moves away from the $AD$ edge of the diagram toward point $Z$. As the liquid reaches point $Z$, a tertiary constituent of $\alpha + \gamma + \delta$ commences to form. Now the liquid composition moves away from the $ACD$ face along the line $de$. At the quaternary eutectic point $e$, any remaining liquid freezes isothermally to the quaternary eutectic constituent composed of $\alpha + \beta + \gamma + \delta$. Coring of the primary, secondary, and tertiary constituents is to be expected.

A series of isopleths is presented in Fig. 19-29. As in previous examples, the $B$ content has been taken at 20% for the entire series, and successive diagrams represent constant quantities of $D$ in intervals of 20%. Five-phase equilibrium, being isothermal in quaternary systems, is represented by a horizontal line in the three sections in which it appears. These drawings serve to illustrate a rule for identifying the phases represented in the various areas, a rule that will be increasingly useful as more complex sections are encountered. Every four-phase field is surrounded by three-phase fields (except along the eutectic isotherm) each of which includes three of the four phases of that region. Two-phase fields, similarly, contain only phases that appear in the three-phase fields with which they are in contact. This is an application of the general principle of the succession of fields stated in Chap. 18.

### An Example of Class II Five-phase Equilibrium

The occurrence of five-phase equilibrium of the second kind is illustrated in Fig. 19-30. Four ternary eutectic systems are again produced by the four components of this quaternary system. In distinction with the previous example, however, one of the ternary eutectics ($BCD$) lies at a temperature below that of five-phase equilibrium. Four-phase tie-tetrahedra emerge in succession: from the $ABD$ face $L + \alpha + \beta + \delta$ just above $T_1$, Fig. 19-30$a$; from the $ABC$ face $L + \alpha + \beta + \gamma$ just above $T_2$, Fig. 19-30$b$; and from the $ACD$ face $L + \alpha + \gamma + \delta$ just above $T_3$, Fig. 19-30$c$. These join at $T_4$ to form a class II five-phase tie-hexahedron representing equilibrium among $L$, $\alpha$, $\beta$, $\gamma$, and $\delta$, Fig. 19-30$d$. The hexahedron divides into two new tie-tetrahedra, one of which ($\alpha + \beta + \gamma + \delta$) persists to low temperature and the other terminates in the ternary eutectic reaction plane of the system $BCD$ at a temperature between $T_5$ and $T_6$.

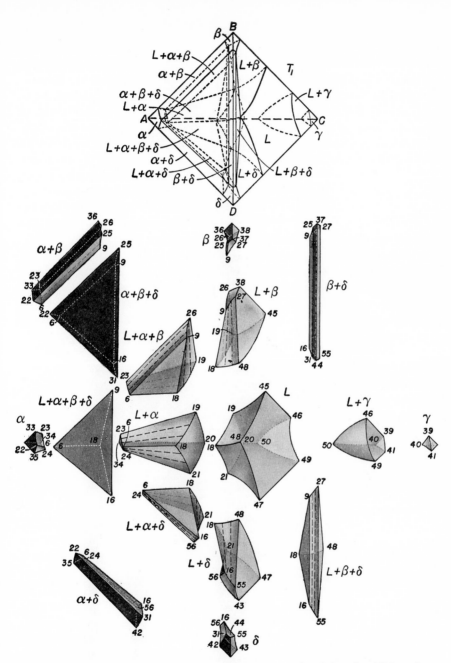

FIG. 19-30a. Six isotherms through a quaternary system involving class II five-phase equilibrium. Four-phase tetrahedra, which originate upon the ternary eutectics of the three systems $ABD$, $ABC$, and $ACD$, join at $T_4$ to form the five-phase hexahedron $L + \alpha + \beta + \gamma + \delta$. Below $T_4$ there are only two four-phase equilibria, of which one terminates upon the ternary eutectic of the system $BCD$ between $T_5$ and $T_6$ and the other ($\alpha + \beta + \gamma + \delta$) persists to low temperature.

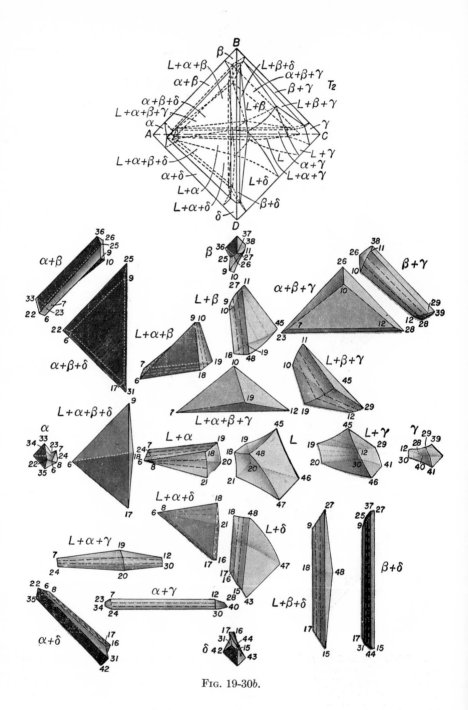

Fig. 19-30b.

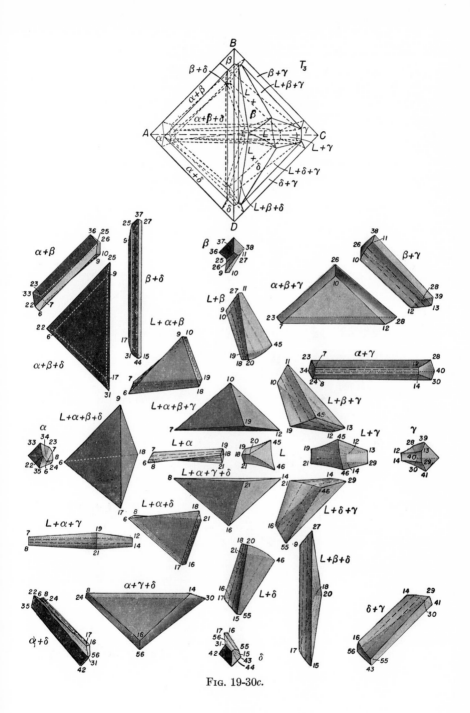

FIG. 19-30c.

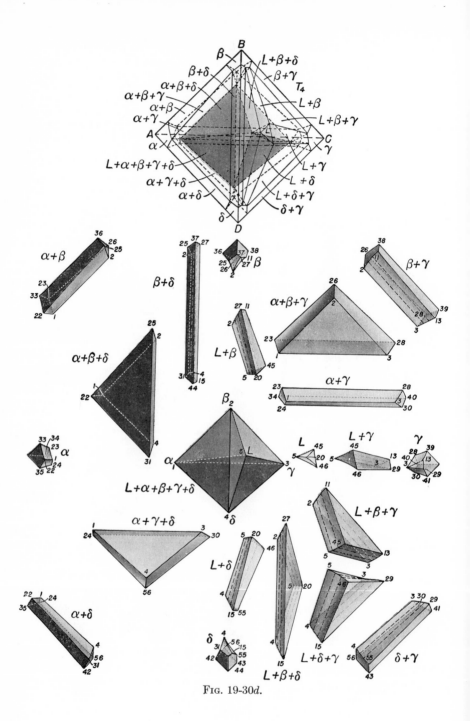

FIG. 19-30d.

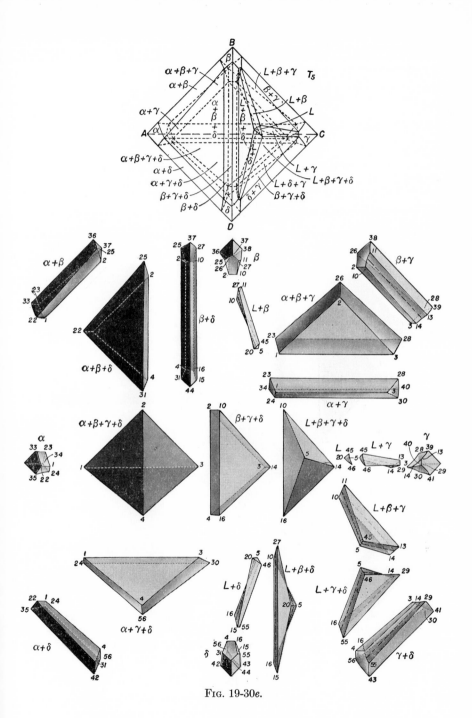

FIG. 19-30e.

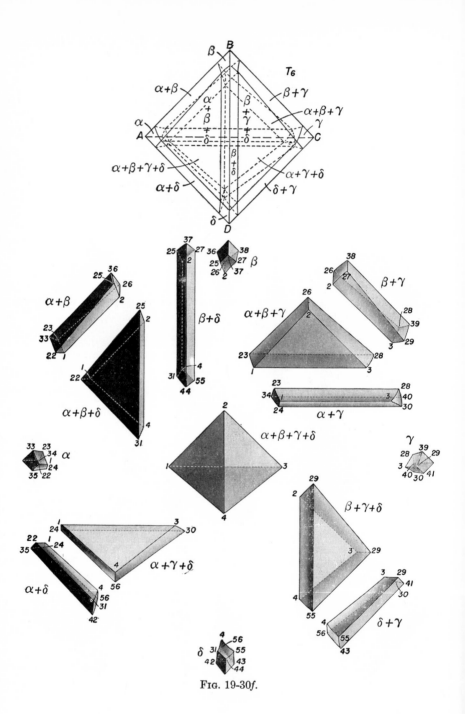

FIG. 19-30f.

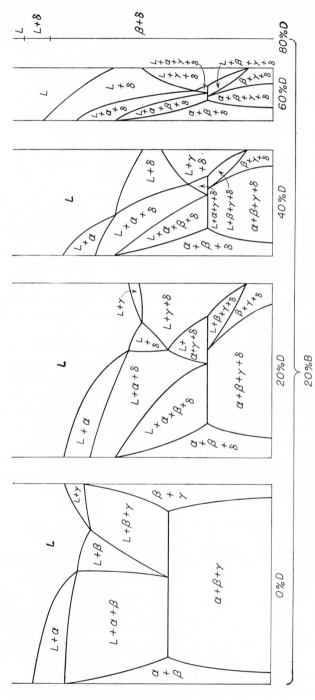

FIG. 19-31. Isopleths through the quaternary system shown in Fig. 19-30.

269

During class II five-phase reaction the quantities of three of the previously existing phases diminish while the other two gain:

$$L + \alpha + \beta \rightarrow \gamma + \delta$$

As with class II four-phase reaction in ternary systems, this type of transformation partakes of both eutectic and peritectic reaction. Hence, it is to be expected that notably incomplete reaction will usually be encountered with any but very slow rates of heating or cooling.

Some examples of two-dimensional isopleths are given in Fig. 19-31. These have been taken in a manner favorable for illustrating the form of the class II five-phase reaction isotherm. A four-phase field involving liquid, $L + \beta + \gamma + \delta$, approaches the reaction plane from the low-temperature side, while three others approach from the high-temperature side although only two of the latter are intersected by these isopleths.

## Other Constructions in Quaternary Systems

Enough examples have now been presented to illustrate the main types of construction that appear in quaternary systems. Five-phase equilibria of the third and fourth kinds, being simply the inverse of the first two, require no detailed discussion. The methods of representing the various kinds of maxima and minima and of quasi-binary and quasi-ternary sections may be deduced with little difficulty by analogy with binary and ternary systems. It is evident that an enormous number of types of quaternary diagrams could be derived by combining the many varieties of binary, ternary, and quaternary equilibria that have been mentioned. Although arduous, the development of the phase diagram for any one of the many cases that have not been discussed should present no unsurmountable problems to one having a clear understanding of the types that have been considered.

## Quinary Equilibrium

With five components the graphical representation of all compositions simultaneously is not possible, since a four-dimensional space would be needed for the purpose. This condition makes the use of isothermal sections impossible and effectively limits quinary diagrams to isopleths in which three of the four concentration variables are held constant. It will be obvious that a prodigious number of such sections would be required to present even a sparse survey of an entire quinary system. Existing surveys have been confined to narrow composition ranges.

From the phase rule it can be seen that a maximum of six phases can exist in equilibrium in a randomly selected quinary isobar. The quinary eutectic is of the form

$$L \rightleftharpoons \alpha + \beta + \gamma + \delta + \eta$$

Five kinds of six-phase isothermal equilibrium are to be expected in

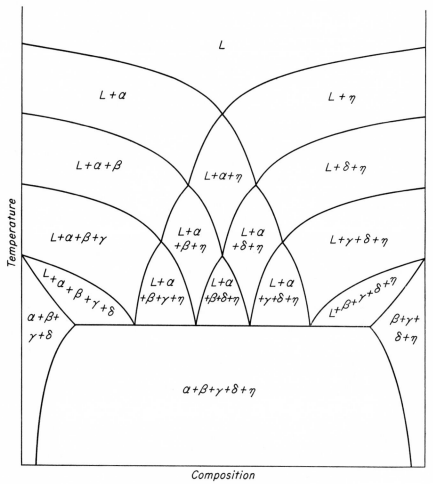

FIG. 19-32. An isopleth through a quinary eutectic system.

quinary systems. The other four are such as

$$L + \alpha \rightleftharpoons \beta + \gamma + \delta + \eta$$
$$L + \alpha + \beta \rightleftharpoons \gamma + \delta + \eta$$
$$L + \alpha + \beta + \gamma \rightleftharpoons \delta + \eta$$
$$L + \alpha + \beta + \gamma + \delta \rightleftharpoons \eta$$

An ideal section through a quinary eutectic diagram is given in Fig. 19-32. It will be seen that an alloy selected at random may freeze in as many as five stages, each producing a microconstituent with an additional phase up to five phases at the quinary eutectic temperature.

## Higher-order Systems

Although practical difficulties make the use of higher-order phase diagrams, in which the composition variable is represented, virtually useless, it is interesting to speculate briefly on the nature of the constitution of the multicomponent systems. It can be seen from the foregoing survey of unary, binary, ternary, and quaternary systems that

1. The number of classes of univariant reaction is always equal to the number of components; there would be 10 classes of isothermal reaction in a decinary system.

2. A region representing $x$ phases must always be bounded by regions representing $x - 1$ and $x + 1$ phases.

3. Other than the pattern of coring and the limits of solubility, no essential structural differences are produced in one-phase alloys as the number of components is increased or in two-phase alloys, and so on.

In general, multicomponent alloys that are composed predominantly of two components, not having the other components present in sufficient quantity to produce additional phases, may be treated as modified binary alloys. This is, in fact, what is always done in the establishment of binary diagrams, because absolutely pure metals can not be obtained or maintained for constitutional studies.

# CHAPTER 20

# PRESSURE-TEMPERATURE DIAGRAMS

When dealing with systems in which the gas phase must be considered, it is not generally satisfactory to limit observations to a fixed pressure as has been done in the presentation of temperature-composition diagrams. Pressure-temperature-composition (PTX) diagrams, such as those shown

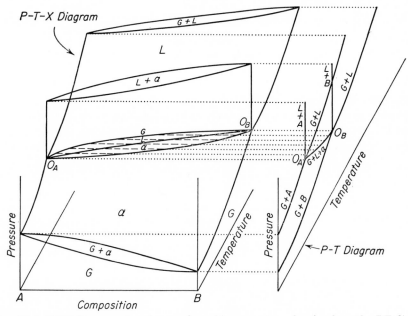

Fig. 20-1. PTX diagram of an isomorphous binary system showing how the PT diagram is derived by projecting all univariant and invariant equilibria upon the PT plane.

in Figs. 3-1 and 4-2, would be ideal for representing equilibria involving the gas phase were it not for the practical difficulty of handling three-dimensional diagrams. As an alternative it is often satisfactory to use pressure-temperature (PT) diagrams, in which the composition variable is ignored. These have the special advantage that they may be applied to systems of any number of components.

The PT diagram of a binary system can be derived from the PTX space diagram by projecting, upon a pressure-temperature plane, all points and lines representing invariant and univariant equilibrium in the space diagram (see Fig. 20-1). All lines from the one-component faces of the space

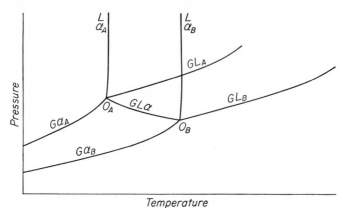

FIG. 20-2. PT diagram of an isomorphous binary system, derived in Fig. 20-1.

diagram are transferred directly to the PT projection, because these are univariant. From the inside, or binary, portion of the space diagram, only tie-lines that connect three or more conjugate phases are transferred to the PT diagram, three-phase equilibrium being univariant in binary systems. Since all tie-lines are both isobaric and isothermal, these project as

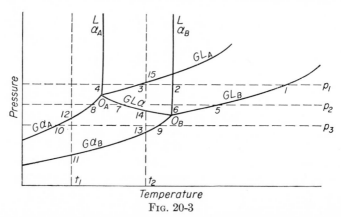

FIG. 20-3

points upon the PT plane. Any surface of the space figure that is generated by tie-lines therefore becomes a line in the PT diagram. In Fig. 20-1 there is a "ruled" surface, between points $O_A$ and $O_B$, composed of tie-lines connecting conjugate compositions of gas, liquid, and solid. This appears on the PT diagram as the line $G + L + \alpha$.

Displayed as normally drawn in two-dimensions, this PT diagram representing binary isomorphous equilib-
ria appears in Fig. 20-2. Each line
is labeled according to the equilibrium represented, the unary lines
representing two-phase equilibrium
and the binary line three-phase equilibrium. The areas between the lines
cannot be identified uniquely, since
the several bivariant and tervariant
equilibria overlap one another in
these areas. Invariant equilibrium
occurs in this example only at the
unary triple points $O_A$ and $O_B$.

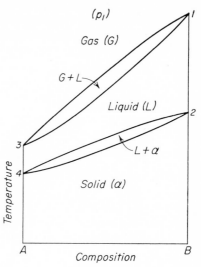

This diagram may be employed to
find the phase changes that occur
during heating or cooling at any desired pressure or the phase changes
that accompany pressure rise or fall

Fig. 20-4. Temperature-composition
section at pressure $p_1$, derived from the
PT diagram of Fig. 20-3.

at a stated temperature. The method
used is illustrated in Fig. 20-3. Sup-
pose that the pressure is fixed at $p_1$. Upon cooling from a very high tem-

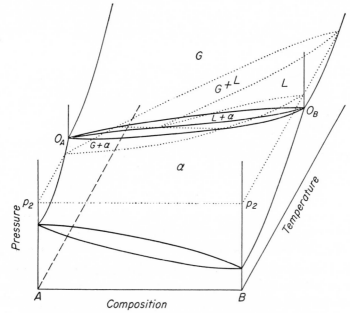

Fig. 20-5. Derivation of the TX section given in Fig. 20-7.

perature, component $B$ will begin to condense to liquid at the temperature of point 1; condensation of all compositions from $B$ to $A$ will occur between 1 and 3. At temperature 2 pure $B$ will begin to freeze, and mixtures of $A$ and $B$ will freeze between 2 and 4. Below the latter temperature all alloys are solid. This description is qualitatively the same as that

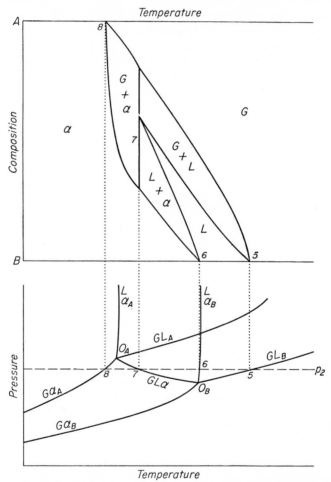

Fig. 20-6. TX section, at $p_2$, shown by dotted lines in Fig. 20-5, derived directly from the PT diagram by reading the temperatures of univariant equilibria where the dashed line $p_2$ intersects solid lines in the PT diagram.

presented by any binary TX diagram of an isomorphous system, such as that given in Fig. 20-4. Only a statement of the compositions of the conjugate liquid and solid phases at each temperature between 2 and 4 and of the conjugate gas and liquid phases between 1 and 3 is lacking.

At a lower pressure $p_2$, Fig. 20-3, the univariant equilibrium among gas, liquid, and $\alpha$ is encountered. Condensation of the vapor phase begins at the temperature of point 5 where pure $B$ condenses to its liquid. From 5 to 7 the condensation of alloys of increasing $A$ content begins. Pure $B$ freezes at the temperature of point 6, and alloys of advancing $A$ content begin to freeze at progressively lower temperatures down to 7. In alloys of intermediate composition the liquid phase decomposes $(L \rightarrow G + \alpha)$ at temperature 7. Below this temperature only solid and vapor can exist at this pressure. The gas now condenses directly to the solid phase at temperatures that decrease to 8 as the composition approaches pure $A$.

These relationships will be more easily perceived in the TX section drawn at pressure $p_2$, Fig. 20-7. Its derivation from the PTX diagram is demonstrated in Fig. 20-5, and its derivation from the PT diagram in

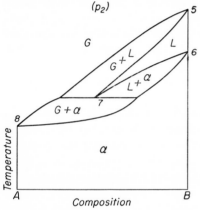

FIG. 20-7. TX section at $p_2$, derived from Figs. 20-3, 20-5 and 20-6.

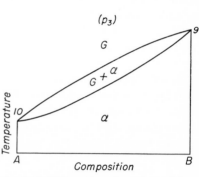

FIG. 20-8. TX section at $p_3$, derived from the PT diagram of Fig. 20-3.

Fig. 20-6. Each line in the PT diagram, Fig. 20-6, intersected by the horizontal line $p_2$ designates the temperature of a univariant equilibrium in the TX diagram. The composition at which each univariant equilibrium should be recorded in the TX diagram is inferred from the labeling of the lines in the PT diagram. All two-phase equilibria are unary and so are plotted at the extreme right or left of the TX diagram, points 5, 6, and 8. Three-phase equilibria are binary and must be represented by a single three-phase tie-line, point 7 in the PT diagram and line 7 in the TX diagram. Three points on the horizontal line arbitrarily chosen to designate the compositions of the three conjugate phases $G$, $L$, and $\alpha$ are connected to corresponding unary points to complete the TX diagram. In Fig. 20-7 it will be observed that the liquid phase undergoes a eutectic type of transformation, in which gas and one solid phase are produced simultaneously (see Table 2, case $e$). In most isomorphous alloys this reaction would occur at such low pressure that it could not be realized except at the surface of the metal, even in the highest vacuum. This is because the hydrostatic head of metal would exceed $p_2$ except at the extreme surface. By this type of reaction, however, the surface of the alloy, frozen in vacuum, would become depleted with respect to the more volatile component $A$. Some other types of systems in which this kind of reaction occurs at higher pressure will be discussed presently. Where surface behavior is of interest, of course, such low pressure transformations become significant.

At still lower pressure, $p_3$ in Fig. 20-3, the liquid phase is missed altogether and alloys of any composition, upon equilibrium heating, sublime directly to the vapor phase or, upon equilibrium cooling, condense directly to the solid phase (see Fig. 20-8). Again, such effects, upon heating, are likely to be confined to the extreme surfaces of pieces of metal, where the low pressure can be attained. Condensation from the vapor phase produces very minor quantities of solid metal at the very low pressures at which such effects usually occur. All such effects are perceptible, however, under special conditions.

Two particularly interesting characteristics of the PT diagram are revealed in the foregoing discussion. First, the general form of the PT diagram can be established upon the basis of a very few facts concerning the alloy system. In this case, it is necessary to know only the approximate coordinates of the triple points of $A$ and $B$ and that the system is isomorphous, without complications. Nothing more is needed for a qualitatively correct representation, although the exact location of each of the curves would require lengthy and often very difficult laboratory investigation. The second characteristic of special interest is that the form of the TX diagram for any given pressure can be derived directly from the PT diagram. Compositions of phases cannot be ascertained from the PT diagram, of course, but the several phases will occur upon the derived TX diagram in the correct composition sequence. Thus, a TX diagram is obtained which is capable of answering many practical questions and, also, one which can be made roughly quantitative with a very few additional data, such as measurement of the exact temperatures of univariant reaction together with determination of the compositions of the three conjugate phases involved in each reaction.

## Pressure-Composition (PX) Diagrams

Two isothermal sections are indicated in Fig. 20-3 by the vertical dashed lines designated $t_1$ and $t_2$. Pressure-composition (PX) diagrams may be derived in the same way that the TX diagrams are derived, namely, by plotting the (numbered) intersections of the isothermal lines and the univariant curves of the PT diagram upon the coordinates of a PX diagram and connecting these points in the only way possible with boundaries of divariant fields (see Figs. 20-9 and 20-10). The resulting PX diagrams are very similar to the corresponding TX diagrams, with reversal of the relative positions of the several fields. Metallurgical uses of PX diagrams are rare, because conditions involving constant temperature and variable pressure are infrequently encountered. The construction

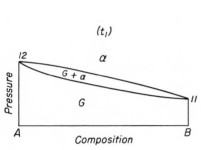

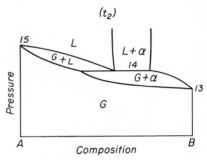

FIG. 20-9. PX section at $t_1$, derived from the PT diagram of Fig. 20-3.

FIG. 20-10. PX section at $t_2$, derived from the PT diagram of Fig. 20-3.

and interpretation of the appropriate PX diagrams parallel so closely those of the TX diagrams, which have formed the principal subject of this book, that no special discussion of the matter is required.

## Invariant Equilibrium in Binary Systems

The PT diagram of a binary eutectic system, corresponding to the PTX diagram shown in Fig. 4-2, is given in Fig. 20-11. As in the previous example, three triple curves of each of the components are indicated, joining at the two triple points $O_A$ and $O_B$. There are, in addition, four binary univariant curves, or *quadruple curves*, $L\alpha\beta$, $G\alpha\beta$, $GL\alpha$, and $GL\beta$, that meet at a *quadruple point* $Q$. This point represents invariant equilibrium among the four phases $G$, $L$, $\alpha$, and $\beta$; its pressure and temperature coordinates are fixed, and the compositions of each of the four conjugate phases are also fixed.

Every quadruple point is associated with four quadruple curves each of which represents equilibrium among a unique combination of three of the four phases occurring at invariant equilibrium. The quadruple curve representing equilibrium among liquid and two solid phases (the eutectic reaction) usually projects almost vertically upward (Fig. 20-11). A slight tilt toward higher temperature with increasing pressure is normal when both components contract upon freezing. The $G\alpha\beta$ curve always proceeds to lower temperature with falling pressure, frequently at a slope such that it would end at zero pressure and the absolute zero of temperature. This curve always lies above both of the unary sublimation curves in eutectic systems, because it represents the total vapor pressure of the two solid phases coexisting. Two quadruple curves $GL\alpha$ and $GL\beta$ connect the quadruple point with the two triple points $O_A$ and $O_B$. One of these proceeds

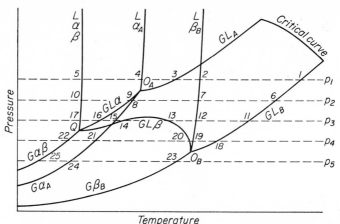

FIG. 20-11. PT diagram of a binary eutectic-type system.

continuously upward from the quadruple point $Q$ to the higher of the two triple points $O_A$.[1] The other of these curves may, and frequently does,

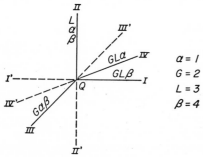

FIG. 20-12. Schematic arrangement of the quadruple curves about the quadruple point $Q$ in Fig. 20-11.

$\alpha = 1$
$G = 2$
$L = 3$
$\beta = 4$

pass through a pressure maximum between the quadruple point and the lower triple point $O_B$ (see Fig. 20-11).

There is a thermodynamic rule that relates the succession of the quadruple curves about the quadruple point with the relative concentrations of the four phases. First, the phases are numbered 1, 2, 3, and 4 in their order of ascending or descending $B$ content; in Fig. 4-2 this order would be $\alpha = 1$, $G = 2$, $L = 3$ and $\beta = 4$. Then the quadruple curves are numbered I, II, III, and IV, using the numeral I to designate that equilibrium from which the first phase is missing, and so on. The quadruple curves, so numbered, should occur clockwise or counterclockwise in the

[1] An exception to this rule occurs when the quadruple point occurs at a pressure maximum. This special case is relatively uncommon and will not be discussed here. For details, see R. Vogel, "Handbuch der Metallphysik," II. Die Heterogenen Gleichgewichte, pp. 199–200, Akademische Verlagsgesellschaft m.b.h., Leipzig, 1937.

following order: I, II′ (i.e., extension of II), III, IV′ (i.e., extension of IV), I′ (i.e., extension of I), II, III′ (i.e., extension of III), and IV (see Fig. 20-12). This rule excludes the possibility of any adjacent quadruple curves meeting at an angle in excess of 180°. By reversing the above computation, the order of compositions of the four conjugate phases represented at any quadruple point as shown on a PT diagram can be ascertained.

Examples of five different configurations of the TX sections, taken at pressure $p_1$ to $p_5$, Fig. 20-11, appear in Figs. 20-13 to 20-17 inclusive. The first of these, corresponding to $p_1$, is the familiar simple eutectic diagram.

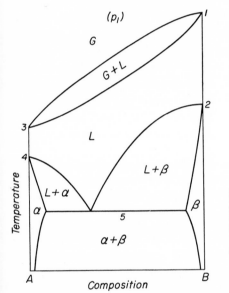

FIG. 20-13. TX section at $p_1$, derived from the PT diagram of Fig. 20-11.

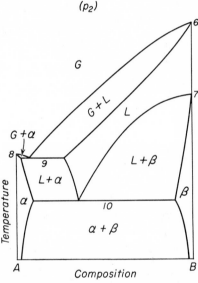

FIG. 20-14. TX section at $p_2$, derived from the PT diagram of Fig. 20-11.

Below the triple point of component $A$, at $P_2$, the liquid phase is confined to $B$-rich alloys, while the $A$-rich alloys cannot be fully melted under equilibrium conditions because sublimation intervenes (Fig. 20-14).

At somewhat lower pressure, $p_3$, the same horizontal line intersects the $GL\beta$ quadruple curve in two places, below its maximum. This gives rise to a double occurrence of the equilibrium $GL\beta$ in the TX diagram (see Fig. 20-15). Accordingly, alloys rich in the $B$ component exhibit double melting and double boiling. Melting first occurs at the equilibrium $L\alpha\beta$ (line 17); this liquid "boils," or actually decomposes into gas and $\beta$, at the lower $GL\beta$ line (line 14). Again, at the upper $GL\beta$ equilibrium (line 13) liquid appears (melting) and finally boils away, incongruently, to vapor. An easily visualized example of this case is found in salt-water mixtures, which are composed of ice and crystalline salt at low temperature, of water and crystalline salt at room temperature, of steam and crystalline salt above the "boiling point," of molten salt at very high temperature,

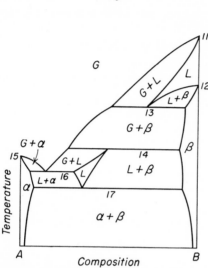

and of vapor alone at extremely high temperature. Similar conditions are to be expected in metal-hydrogen systems where no hydride forms. Let $A$ represent hydrogen and $B$ the metal; the upper liquid field (Fig. 20-15) would represent the molten metal with dissolved hydrogen, and the lower liquid field, which would occur only at very low temperature and close to the "$A$ side" of the diagram, would represent liquid hydrogen. If molten metal containing dissolved hydrogen were cooled, the primary freezing process would end with an evolution of gas which would probably be nearly pure hydrogen because the $GL\beta$ line would extend almost across the diagram. Remelting at lower temperature would not be

Fig. 20-15. TX section at $p_3$, derived from the PT diagram of Fig. 20-11.

observed, because room temperature occurs far above the lower boiling point in this case.

Somewhat similar conditions are found in the TX section at $p_4$, Fig. 20-16, below the quadruple point. No low-temperature occurrence of the

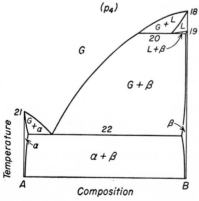

Fig. 20-16. TX section at $p_4$, derived from the PT diagram of Fig. 20-11.

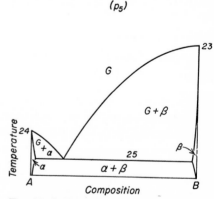

Fig. 20-17. TX section at $p_5$, derived from the PT diagram of Fig. 20-11.

liquid phase is found, however, at this low pressure. Alloys rich in $B$ would exhibit gas evolution during freezing, as at higher pressure. Finally, at very low pressure, $p_5$, Fig. 20-17, the liquid phase is entirely absent from "equilibrium transformation." Upon cooling, the gas phase itself deposits a two-phase alloy of $\alpha + \beta$. This type of reaction is usually confined to such low pressure in metal systems that familiar examples are lacking, although a closely related behavior is found in the operation of arsenic kitchens when oxides of impurity metals condense with the arsenic.

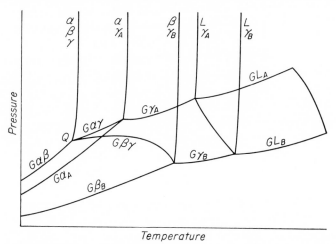

Fig. 20-18. PT diagram of a system involving a binary eutectoid-type equilibrium at $Q$.

By changing the pressure and temperature relationships on the PT diagram, some other configurations of the TX diagram are obtained. However, the previously cited examples include all cases that are well known in metal systems, and since, furthermore, the other TX sections are readily derived, there is little to be gained by exhausting the subject here.

## Some Other Binary Quadruple Points

All quadruple points associated with high-pressure equilibria of the eutectic class, i.e., eutectoids, monotectics, have the arrangement of quadruple curves illustrated in Fig. 20-12, namely, one "vertical" curve, one proceeding toward lower temperature, and two toward higher temperature. An example of a PT diagram from which the simple eutectoid diagram of Fig. 5-1 may be derived is given in Fig. 20-18. Here the quadruple point $Q$ represents equilibrium among gas, $\alpha$, $\beta$, and $\gamma$. It will be evident that an analysis of the low-pressure behavior of this system will lead to the conclusion that the decomposition of the $\gamma$ phase, like that of the liquid phase in the previous example, can be accompanied by gas evolution.

The monotectic system, illustrated in Fig. 6-1, may be derived from the PT diagram of Fig. 20-19. Two quadruple points appear in this diagram, $Q_1$ corresponding to eutectic reaction at low temperature and $Q_2$ to

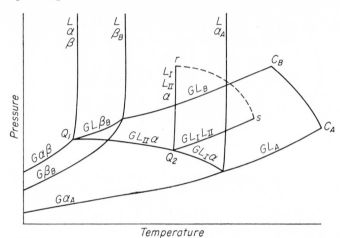

FIG. 20-19. PT diagram of a system involving a binary monotectic-type equilibrium at $Q_2$.

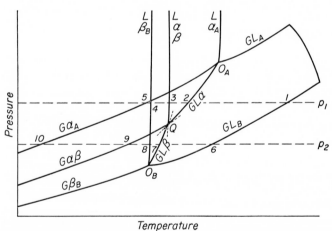

FIG. 20-20. PT diagram of a system involving a binary peritectic type equilibrium at $Q$.

monotectic reaction at higher temperature. Three phases, gas, liquid II, and $\alpha$, are common to the two quadruple points, and the quadruple curve representing equilibrium among these three phases serves to connect the quadruple points. The $GL_IL_{II}$ curve is shown ending in a critical point at $s$; other configurations are possible. In like manner the $L_IL_{II}\alpha$

curve is shown terminating upon a critical point at $r$. When this happens the field of $L_I + L_{II}$ has a critical curve following the path of the dashed line $rs$.

With peritectic reaction, as in Fig. 8-1, the disposition of the quadruple curves is reversed: one is "vertical," two proceed toward lower tempera-

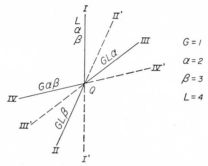

FIG. 20-21. Schematic representation of the arrangement of the quadruple curves about the quadruple point $Q$ in Fig. 20-20.

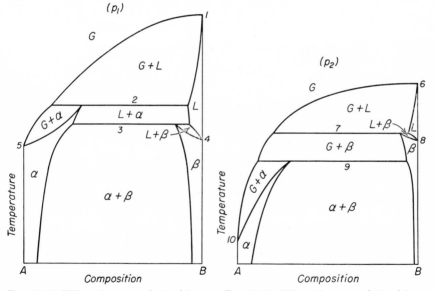

FIG. 20-22. TX section at $p_1$, derived from the PT diagram of Fig. 20-20.

FIG. 20-23. TX section at $p_2$, derived from the PT diagram of Fig. 20-20.

ture, and only one toward higher temperature (see Fig. 20-20). The application of the rule of succession of quadruple curves is illustrated for the peritectic case in Fig. 20-21. All three-phase equilibria of the peritectic class, such as peritectoids and syntectics, originate upon quadruple points of the same form. Two low-pressure configurations of the TX dia-

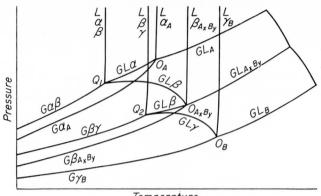

FIG. 20-24. PT diagram of a system having a fully congruent intermediate phase $AxBy$.

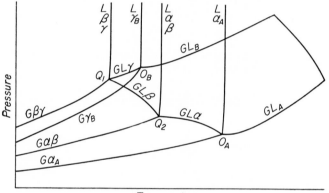

FIG. 20-25. PT diagram of a system having a peritectic-type binary equilibrium $Q_2$ and a eutectic-type binary equilibrium $Q_1$.

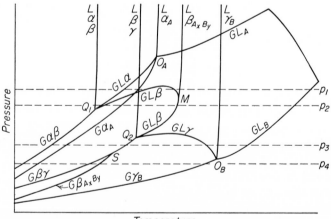

FIG. 20-26. PT diagram of a system having an intermediate phase that is incongruent in the pressure-temperature range from the maximum sublimation point $S$ to the minimum melting point $M$.

gram, characteristic of this type of system, are shown in Figs. 20-22 and 20-23. The "peritectic" decomposition of a solid phase into gas and liquid appears in both of these TX sections. A reaction similar to the upper three-phase equilibrium in Fig. 20-23 is thought to occur in the palladium-hydrogen system where the palladium-rich solid phase melts with evolution of gas.

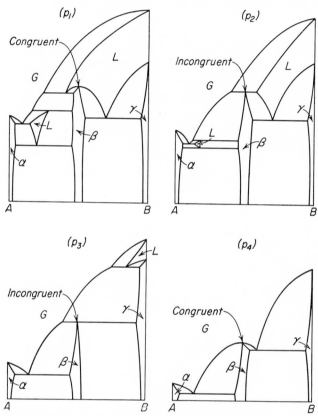

FIG. 20-27. Temperature-concentration sections derived from the PT diagram of Fig. 20-26.

Intermediate phases may be congruent throughout the entire range of pressure and temperature, having their own triple points and triple curves as shown in Fig. 20-24; they may be incongruent in all ranges as is the $\beta$ phase of Fig. 20-25 (Fig. 8-2 is derived from this PT diagram); or the intermediate phase may be congruent in some ranges and incongruent in others. The latter case, as illustrated in Fig. 20-26, is thought to occur in several metal-oxygen systems, notably copper-oxygen. Instead of having a triple point, the intermediate phase is congruent only above a certain minimum pressure, the minimum melting point $M$. Below this pressure it melts with decomposition and sublimes, also with decomposition, down to the pressure of the maximum sublimation point $S$, below which congruent sublimation is observed. These conditions are demonstrated in the isobaric sections of Fig. 20-27, taken at the four pressure levels indicated in Fig. 20-26. The maximum sublimation point may occur anywhere upon the curve $G\beta\gamma$ or anywhere below $M$ on the curve $GL\beta$.

## Ternary and Higher-Order PT Diagrams

Since the concentration variable is not represented in PT diagrams, there is fundamentally no limit to the number of components that may be handled. Actually, the diagrams become extraordinarily intricate, and

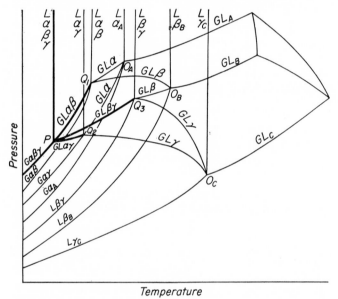

Fig. 20-28. PT diagram of a ternary eutectic-type (class I) system.

therefore difficult to read, as the number of components is increased. Even a simple ternary eutectic system (Fig. 20-28) is relatively complex in its representation. In this example there are three triple points and sets of triple curves, corresponding to the three components, three binary (eutectic) quadruple points and accompanying quadruple curves, and one ternary quintuple point $P$ from which issue five quintuple curves, three of which connect the quintuple point with the three quadruple points. All three of these quintuple curves proceed from the quintuple point to higher temperature, while one curve ($G\alpha\beta\gamma$) goes to lower temperature.

This is characteristic of all class I ternary equilibria (Fig. 20-29a and recall Fig. 18-1i). Quintuple points representing class II equilibrium have two curves running toward higher temperature and two toward lower temperature (Fig. 20-29b), while class III (Fig. 20-29c) equilibrium is the reverse of class I (recall Fig. 18-1i, j, and k).

Complex PT diagrams are read in the same manner as are the binary types. It is not difficult to derive ternary temperature-concentration diagrams as can be seen by comparing Fig. 20-28, at high pressure, with Fig. 14-1, to which it corresponds.

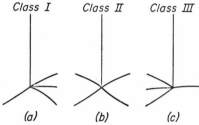

FIG. 20-29. Schematic arrangement of the quintuple curves about the quintuple point (such as P in Fig. 20-28) for the three classes of ternary univariant equilibrium.

It will be evident that higher-order systems will involve increasing numbers of phases in their univariant and invariant equilibria. Quaternary systems will have four kinds of sextuple points each with six sextuple curves, quinary systems five kinds of septuple points with seven radiating septuple curves, and so on. No complex metal systems are known to have been investigated with regard to their pressure-temperature equilibria, and no diagrams are available.

CHAPTER 21

# DETERMINATION OF PHASE DIAGRAMS

The modest collection of phase diagrams of metal systems that is now available to us is the product of the painstaking labor of a very large number of skillful investigators working in all parts of the world during the past three-quarters of a century. It is a constantly growing body of literature, improving both in the extent of its coverage and in the precision of its content. Few, if any, diagrams may be considered complete and final. Repeated investigation, with refinements in apparatus and techniques, leads to their frequent revision. Much of this refinement has been achieved through the modification and adaptation of a few basic research methods to fit the special requirements of specific alloy systems. It is not feasible, therefore, to present a collection of the "best" methods of investigation (no method is "best" for all systems), nor is it feasible to give a standardized procedure for any one method. It is possible, however, to outline the basic methods from which most of the specialized techniques are derived and to state those principles of investigation which are common to all cases.[1] For detailed guidance the investigator must, of necessity, turn to the research literature, because the specialized problems met in constitutional studies are almost as numerous as are the alloy systems themselves.

In the determination of temperature-composition diagrams two kinds of information are usually sought, namely, (1) the general plan of the diagram, including the number, disposition, and identity of the phases and of the univariant equilibria in which they partake, and (2) the temperature and compositions along all boundary lines (and surfaces), particularly at junctions of boundaries. Some research methods are suited to the making of general surveys, others to the measurement of conditions in specific equilibria, and yet others are useful in both categories. Thus, a diffusion couple may sometimes be used to find the number and order of

---

[1] A detailed outline of experimental procedures is given by W. Hume-Rothery, J. W. Christian, and W. B. Pearson, "Metallurgical Equilibrium Diagrams," The Institute of Physics, London, 1952. Some helpful experimental techniques are also described by A. U. Seybolt and J. E. Burke, "Experimental Metallurgy," John Wiley & Sons, Inc., New York, 1953.

the phases in a system; metallographic or X-ray methods may be employed to locate the boundaries between fields involving only solid phases; while thermal analysis is capable of locating the liquidus lines with precision and at the same time indicating the general disposition of phases and univariant equilibria in the system.

For the exact location of boundary lines it is necessary to establish equilibrium conditions and at the same time to have an accurate measure of the temperature and of the chemical composition of the material under observation. Attention has been directed in previous chapters to the fact that true equilibrium conditions are rarely attainable within a reasonable length of time. To overcome this difficulty it is usually possible to approach the equilibrium from two directions, thereby establishing limits between which the true equilibrium must occur. The longer the time available for the approach to equilibrium from each direction, the narrower will be the range of uncertainty. For example, the liquidus temperature of an alloy, as observed during cooling, is usually depressed somewhat below its true value, while an observation made during heating will give a result that is a little too high. If the rates of heating and cooling are decreased, the overheating and undercooling will be less marked and the range of uncertainty will diminish. This is a basic principle which may be, and wherever possible should be, applied to all measurements of phase equilibria.

Temperature measurement is a subject in itself, involving many problems yet to be solved. There are several excellent books on pyrometry[1] that treat the subject from the viewpoint of accuracy of temperature measurement. In the selection of methods, however, it is important to consider also the possibility of disturbing the equilibrium under observation in the course of temperature measurement. Any apparatus that comes in contact with the research sample must be inert at the temperature concerned, so that no substantial contamination will occur to vitiate the observations. This requirement becomes difficult to meet where the temperatures are very high or the metals are highly reactive and accounts for much of the uncertainty concerning the constitution of high-melting-alloy systems. It is important, in addition, that the temperature-measuring element be in good thermal contact with the specimen and that it be not so large in comparison with the sample that it carries away a significant quantity of heat.

Accurate chemical analysis is of extreme importance in the determination of phase diagrams. There are, of course, many excellent reference

[1] See, for example, W. P. Wood and J. M. Cork, "Pyrometry," McGraw-Hill Book Company, Inc., New York, 1941; Pyrometric Practice, *Natl. Bur. Standards Tech. Papers* 170; R. B. Sosman, "Pyrometry of Solids and Surfaces," American Society for Metals, Cleveland, 1940.

books dealing with analytical methods,[1] but in the study of alloy systems not previously investigated it is frequently found that well-standardized methods of chemical analysis are lacking. This creates the necessity of devising and standardizing new procedures, a task which is often more arduous than the constitutional study itself. Wherever possible it is best to put this part of the work in the hands of a skilled analytical chemist. Unreliable chemical analyses are often worse than none at all, because analytical errors may become very large, exceeding the errors to be expected in the computation of composition from a record of the materials introduced into the alloy, the "synthetic analysis." When there is divergence between the chemical and synthetic analyses, the "commonsense" approach is to be recommended. Trust should be placed in the chemical analysis if it has been obtained by the skilled application of a well-standardized method; otherwise, unless there is reason to suspect composition change during the production of the alloy sample, greater trust should be placed in the synthetic analysis. In any case, however, a lack of agreement between the chemical and synthetic analyses must be regarded as an uncertainty, detracting from the acceptability of the constitutional study.

Regardless of the analytical method or the skill with which it is employed, the result can be no more representative of the composition of the alloy investigated than is the sample taken for analysis. Segregation can easily cause a variation of several per cent from point to point in castings of some materials. The analytical sample should be taken in such a way as to obtain an average (or representative) analysis. Drastic working of the metal, alternating with homogenizing heat treatments, will sometimes eliminate composition differences and simplify the problem of sampling. It is sometimes found that the composition of the alloy changes in the course of remelting or during heat treatment, owing to the selective oxidation or vaporization of one or more of its components. Measures should, of course, be taken to avoid composition variations by the adjustment of the conditions of the experiment, but when this cannot be done, repeated sampling and analysis throughout the course of the constitutional study become necessary.

Of equal importance is the maintenance of purity and initial composition of the material that is being investigated. It goes without saying that only the purest obtainable metals should be used in constitutional studies. Just what limits of impurity are tolerable can be ascertained only by experience with the specific alloy system; sometimes, as little as 0.001% of a certain impurity becomes significant while much larger quantities of

[1] See, for example, "Scott's Standard Methods of Chemical Analysis," D. Van Nostrand Company, Inc., New York, 1939; F. P. Treadwell and W. T. Hall, "Analytical Chemistry," John Wiley & Sons, Inc., New York, 1924.

other impurities produce no detectable effects. The use of high-purity metals is of little avail, however, unless the purity is maintained in the production and manipulation of the experimental alloys.

Anything that comes in contact with the samples, particularly when the metals are molten, is a possible source of contamination. Crucibles should be selected with great care, using only materials that are insoluble in and nonreactive with the alloy. Evidence of attack of the crucible by the alloy is usually grounds for rejecting the experimental results, but the absence of such evidence does not guarantee the absence of contamination. When suitable crucibles are unobtainable, it is often possible to coat ordinary crucibles with an inert material. Failing this, it is sometimes possible, as a last resort, to use the experimental alloy as its own "crucible" (at considerable sacrifice of temperature control) by melting a pool in its top surface with a fine torch or other concentrated source of radiated heat. For constitutional studies at temperatures below the melting point, alloys that cannot be maintained in a satisfactory state of purity, if made by melting, or that cannot be melted to a homogeneous liquid can sometimes be produced by mixing the component metals in powder form, compressing to a dense briquette, and heat treating for a long time to permit alloying by diffusion in the solid state.

Contamination from the atmosphere is easily overlooked and is often very damaging. In extreme cases the liquidus temperature has been observed to be depressed several hundred degrees simply as a result of the absorption of the constituents of air. Usually, the effects are much smaller and are confined to the selective oxidation of one or more of the components of the alloy, thereby altering its composition. Protection from gaseous contamination may sometimes be achieved by the use of flux covers, such as borax for materials of intermediate-melting-point or oil for low-melting-point metals, or by the use of inert gas atmospheres, argon and helium usually being satisfactory. Vacuum melting and heat treatment are hazardous from the standpoint of maintenance of composition unless the vapor pressures of all components concerned are very low indeed. Vapor losses may be minimized by the use of positive pressures of inert gases or suitable flux covers.

Before turning to a discussion of the basic research methods, a few words concerning the confirmation of laboratory findings are in order. No phase diagram can be considered fully reliable until corroborating observations have been made by at least two independent methods. Even so, the diagram cannot be accepted if its construction violates the phase rule or any of the other rules of construction that have been derived by thermodynamic reasoning. If violations of these rules are encountered, it may be concluded with assurance that the experimental observations or their interpretation are at fault.

## Thermal Analysis

By far the most widely used method of constitutional investigation is thermal analysis. Its principle is extremely simple. If the temperature of a

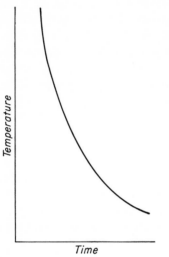

FIG. 21-1. Cooling curve with no phase change.

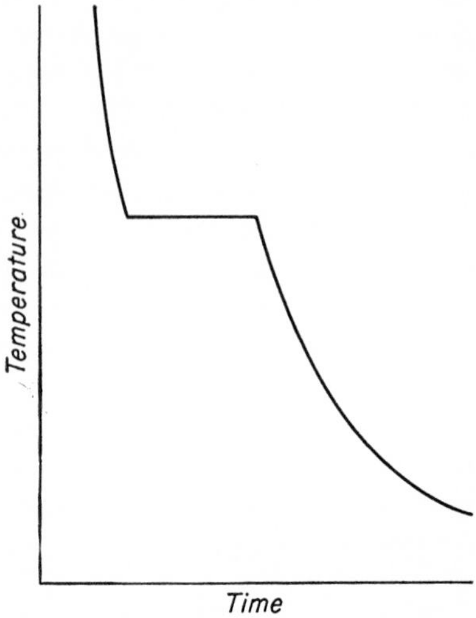

FIG. 21-2. Ideal freezing curve of a pure metal.

body which undergoes no phase change is determined as a function of time while cooling through a temperature interval, the resulting *cooling curve* will be a smooth line (see Fig. 21-1). The occurrence of a phase change that is accompanied by the evolution of heat, such, for example, as the heat of crystallization from the melt, will cause a delay in cooling. Thus, a pure metal, in passing through its freezing point, suffers a *temperature arrest* (Fig. 21-2); the temperature is maintained near the freezing point by the latent heat evolved until freezing is done, whereupon normal cooling is resumed.

As has been mentioned previously, undercooling commonly precedes the onset of freezing. The cooling curve follows the path of natural cooling past the equilibrium freezing temperature (see Fig. 21-3, cooling curve). When crystallization finally

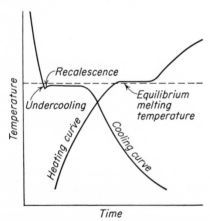

FIG. 21-3. Natural cooling and heating curves of a pure metal.

begins, it often proceeds so rapidly that the evolution of latent heat raises the temperature again almost to the equilibrium freezing point. This temperature rise is called *recalescence*. At the end of the thermal arrest the

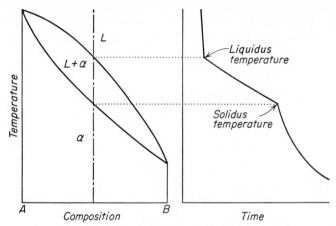

Fig. 21-4. Ideal freezing curve of a solid-solution alloy.

transition to the normal cooling rate is not always abrupt. A "rounding" of this portion of the curve may be associated with the presence of impurities in the metal or with an improper arrangement of the experimental apparatus or both. Upon heating, melting is preceded by a very slight overheating and the thermal arrest corresponding to melting lies very slightly above the equilibrium melting temperature (Fig. 21-3, heating curve). The temperature range within which equilibrium freezing occurs must lie between the horizontal portions of the cooling and heating curves. If the rates of cooling and heating are very slow, perhaps 0.5 to 0.1°C per minute, this temperature range will usually be reduced to a fraction of 1°.

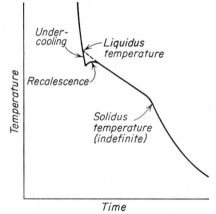

Fig. 21-5. Natural freezing curve of a solid-solution alloy.

Bivariant transformations, such as the freezing of solid-solution alloys, give rise to a change in the slope of the cooling or heating curve (Fig. 21-4), i.e., a retardation in the cooling or heating rate rather than a thermal arrest. Undercooling and coring effects modify this type of cooling curve greatly (Fig. 21-5). Recalescence following the initial undercooling may cause the curve to return almost to its "ideal" path so that the "true" liquidus temperature may be estimated by extrapolating the nonisothermal arrest back to an intersection with the initial limb of the cooling curve. This is never an entirely satisfactory procedure, but the estimated liquidus point constitutes a closer approximation of the true liquidus temperature than does the first recorded break in the cooling curve. The return of the slope of the cooling curve to that of normal cooling, marking the solidus temperature, is hardly ever distinct, owing to the interference of the coring effect. Other methods must usually be employed to locate the solidus, although heating curves, run on previously homogenized samples, give more distinct solidus indications than do cooling curves.

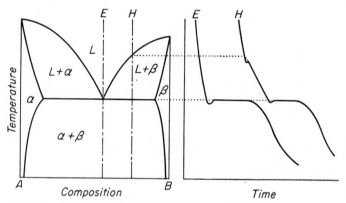

FIG. 21-6. Natural freezing curves of eutectic $E$ and hypereutectic $H$ alloys.

Univariant equilibrium, being isothermal, should be expected to produce a horizontal arrest. Curve $E$ of Fig. 21-6 is characteristic of cooling curves of eutectic alloys. Undercooling and recalescence are often very pronounced; the horizontal portion of the curve lies slightly below the true eutectic temperature and usually merges gradually with the final limb of the curve. Hypo- and hypereutectic alloys yield cooling curves that combine the characteristics of the solid-solution and eutectic-type curves. Under comparable circumstances the eutectic arrest in such alloys is always of shorter duration than in the eutectic alloy itself, because there is a smaller fraction of eutectic liquid remaining at the eutectic temperature. With peritectic alloys the isothermal arrest is often very short indeed and the "rounded" portion, in which the arrest merges into natural cooling, is extended both in time and in temperature range. Similar forms of cooling curves are found with all the other types of univariant transformation. Undercooling is normally more pronounced in eutectoid transformation than with eutectic or monotectic reactions and tends to be even greater in all transformations of the peritectic class. Bivariant phase changes in the solid state (solvus curves, for example) rarely give perceptible indications on cooling or heating curves; other methods must usually be employed for their determination.

Ternary and higher order alloys give more complex cooling curves. A random alloy of a ternary eutectic system, for example, produces a curve with two consecutive thermal retardations, corresponding to primary (or tervariant) crystallization and secondary (or bivariant) crystallization, and one isothermal arrest corresponding to eutectic decomposition (Fig. 21-7).

Thus, it is seen that thermal analysis is useful in the location of liquidus curves and of most isothermal reactions. It fails where the tendency toward undercooling or overheating is excessive or where the heat evolution or absorption associated with a phase change is small. An alternate procedure known as *dip sampling*, which will be described later, can usually be applied to the determination of liquidus curves where undercooling is serious. Several more sensitive but less general techniques are available for locating univariant reactions.

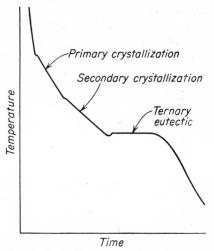

FIG. 21-7. Cooling curve of an alloy of a ternary eutectic system.

Thermal analysis is often employed for the preliminary survey of alloy systems. Cooling curves taken with samples in a continuous series of compositions across the alloy system give a rough approximation of the complete liquidus, the course of which indicates the presence of eutectics, peritectics, and congruently melting intermediate phases. At the same time, most or all of the isothermal reactions occurring in the solid state will be indicated. Monotectics and syntectics are indicated by liquid immiscibility as well as by thermal arrests.

Of several short cuts that have found use in general survey work, two are worthy of special mention. These are means for locating approximate binary and ternary eutectic compositions with a minimum number of experiments. The first method is illustrated in Fig. 21-8. If it is assumed that the duration of the eutectic arrest is a *linear function* of the composition of the alloy, then hypoeutectic alloys will give arrests of length *proportional* to the vertical distance between the lines *ab* and *ae* at the corresponding compositions. Any two hypoeutectic alloys will serve to define the line *ae*, and any two hypereutectic alloys will define the line *eb*. The intersection of these two lines at point *e* designates the eutectic composition. In this way a minimum of four observations may serve to locate the eutectic point. In practice, the method is found to yield only a very crude approximation of the eutectic composition because of the interference of coring effects.

A ternary version of this method is described by the diagrams of Fig. 21-9. On a suitably located isopleth *ab* the points *x* and *y* are found by means of cooling curves. This requires a minimum of eight experiments if it is assumed that the lines intersecting at *x* and *y* are straight, otherwise

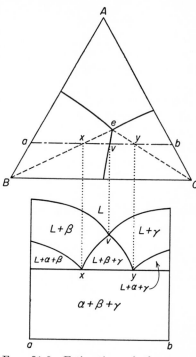

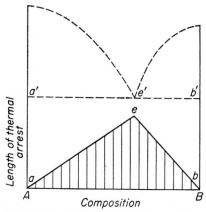

Fig. 21-8. Length of the thermal arrest, in a cooling experiment, as a function of alloy composition in an ideal eutectic system.

Fig. 21-9. Estimation of the ternary eutectic composition by the determination of one isopleth and assuming no solid solubility.

more. If it is asumed, further, that the tie-lines originating upon the eutectic point and passing through *x* and *y* would, if projected, intersect the *B* and *C* corners of the diagram, respectively, then it becomes possible to draw two straight lines *Bxe* and *Cye*. These are the lowest tie-lines of the *L* + *β* and *L* + *γ* fields, and their intersection locates the point *e*. The assumption that these tie-lines should intersect the corners of the ternary diagram is rarely justified, but the use of two parallel isopleths locating two points on each tie-line is capable of giving precise results (Fig. 21-10). The same principle may be applied to the location of quaternary eutectic points, where, upon the same basis, a minimum of 48 individual experiments would be required for a first approximation of the location of the quaternary eutectic point.

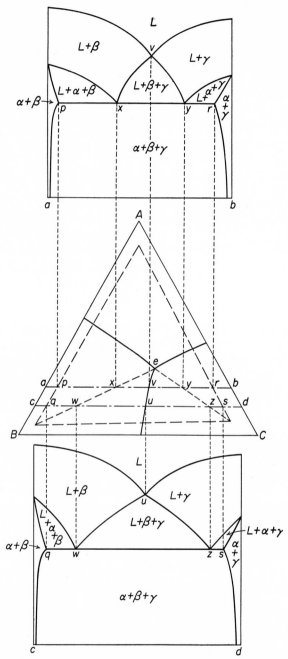

Fig. 21-10. Modification of Fig. 21-9, using two isopleths and making no assumption with respect to the existence or extent of solid solubility.

The experimental arrangements for thermal analysis may be very simple or relatively elaborate. Cooling curves of the low-melting alloys can be made with no more apparatus than a thermometer, a watch with a second hand, and provisions for melting the alloy. When the melt is ready, the heat is turned off and the thermometer secured with its bulb near the center of the sample. Temperature readings are then made at uniform intervals of time, or the time required for each degree (or 5 or 10°) of cooling is noted. These data are then plotted on time-temperature axes to obtain a standard cooling curve. Some investigators prefer to plot the data as an "inverse rate curve," wherein the time required for each degree or stated number of degrees of temperature fall is plotted as a function of temperature. This method tends to accentuate the thermal arrests (see Fig. 21-11).

Refinements of this primitive technique that will serve to minimize undercooling and thereby increase the accuracy of the results include stirring, seeding, and measures to reduce the rate of cooling. The cooling rate may be decreased simply by enclosing the crucible of metal in a thermal insulator. A thick-walled crucible furnace with a good cover will ordinarily cool very slowly in the low-temperature range. Mechanical stirring of the melt until the liquidus temperature has been observed is one of the most effective means available for minimizing the undercooling of metals. Seeding, by the addition of small solid particles of the alloy under investigation shortly before the liquidus temperature is reached, is sometimes helpful, but it is not usually so effective with metals as it is with organic chemical systems.

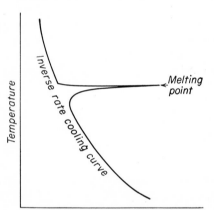

Fig. 21-11. An "inverse-rate" cooling curve.

Glass thermometers, although capable of a high degree of accuracy, are not generally useful in constitutional studies, both because of the very limited temperature range within which they may be employed and because they are subject to damage by the crushing action of the metal as it contracts during freezing. Thermocouples are most commonly used at temperatures up to about 1500°C. There are special thermocouple materials that may be used at temperatures considerably in excess of this, but optical and radiation-type pyrometers are fully as accurate and generally easier to use. Care must always be taken to protect the thermocouple from the liquid metal or from damage by fluxes and special atmospheres.

Where optical or radiation instruments are used, black-body conditions must, as a rule, be maintained because the radiation corrections for very few alloys are known and these are not likely to include the materials under study. It is always desirable to take temperature measurements at the exact center of the sample; this is not absolutely necessary, because the temperature readings will be the same whether the instrument is centered or not, but the sharpness of definition of inflection points on the cooling curve is increased by centering the thermocouple. The use of an automatic recorder for continuous temperature measurement is highly desirable, for the labor of making periodic temperature readings over long periods of time is both exhausting and unnecessary. Great care must be exercised, however, in the frequent calibration of the instrument.

Controlled cooling and heating rates are advantageous in thermal analysis both because they may be employed to evaluate the extent of undercooling and overheating at transformation temperatures and because the cooling curve may be made normally straight with all deviations from the linear path being ascribable to phase changes. Several devices for controlling the rate of temperature change are available. One very useful method which can be arranged with almost any kind of automatic temperature controller has been described by C. S. Smith[1] and is illustrated in Fig. 21-12. The crucible $C$ containing the sample of metal $M$ must be made of a good thermal insulator, such as porous brick lined with alundum cement to prevent leakage. A differential thermocouple is arranged with one junction at the inside surface of the crucible wall $U$ and the other at a corresponding point outside the crucible wall $V$. This thermocouple is connected to an automatic temperature controller which regulates the power input to the furnace windings $W$. With this arrangement the temperature difference between $U$ and $V$ is held constant and the rate of cooling or heating is determined by the rate of conduction of heat across the temperature gradient $UV$. Any desired rate, short of the natural cooling or heating rate of the assembly, can be obtained by adjusting the temperature difference between $U$ and $V$ through the setting of the temperature controller. The cooling or heating curve is traced by the automatic recorder connected to the thermocouple $T$. In the absence of thermal arrests, the curves are nearly linear,[2] and since the rate of heat extraction or input is constant, the apparatus may be calibrated to yield quantitative measurements of latent and specific heats. Precision measurements of this kind require additional precautions

[1] C. S. Smith, A Simple Method of Thermal Analysis Permitting Quantitative Measurements of Specific and Latent Heats, *Trans. Am. Inst. Mining Met. Engrs.*, **137**:236–245 (1940).

[2] Deviation from linearity will result from a variation of the thermal conductivity of the crucible with temperature level.

against irregular heat losses through convection and spurious heat evolution resulting from oxidation and other side reactions. For further details the article by C. S. Smith should be consulted.

Another use of the differential thermocouple in thermal analysis is for determining the exact difference in melting point between two alloys that melt or transform at nearly the same temperature. It will be recalled that in the limiting case, the binary peritectic and eutectic diagrams become identical. Only if the temperature of the three-phase reaction is

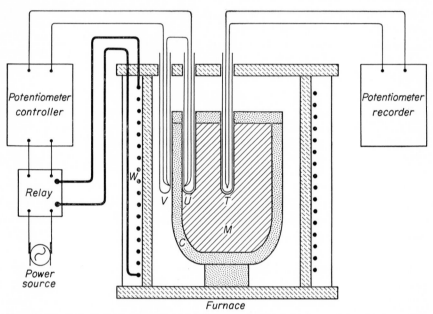

FIG. 21-12. Schematic arrangement of the experimental equipment to obtain a controlled rate of cooling or heating, using only fixed temperature-control equipment.

higher than the melting point of the pure component can it be said that the reaction is peritectic, or if lower, that it is eutectic. One bead of the differential thermocouple placed in the pure metal and the other in the alloy will give a very sensitive indication of the direction and amount of the temperature difference and thus distinguish the two cases.[1]

## Dip Sampling

One of the oldest techniques of constitutional investigation is dip sampling. Liquidus points are determined by holding the sample at a fixed temperature in the liquid plus solid range until equilibrium is established,

[1] Y. Dardel, The Lead Aluminum Diagram, *Light Metals*, **9**:220–222 (1946).

permitting all crystals to separate by settling (or floating) and decanting a sample of the clear liquid for analysis. When working with water solutions the solid can also be separated from the liquid phase by filtering and may then be analyzed separately to establish a point on the solidus curve. The procedure must, of course, be repeated at a series of fixed temperatures to obtain a complete curve.

Dip sampling is awkward when applied to metal systems where the temperatures are high, and the solid phase often grows in a continuous dendritic network that interferes with its separation from the liquid. In certain cases, however, it can be applied and with very definite advantage. Consider for example the hypoeutectic liquidus of the aluminum-silicon system (Fig. 4-5a). The primary crystallization of the $\beta$ phase (silicon-rich), in this case, occurs by the separation of individual idiomorphic particles (Fig. 4-10b). These quickly float to the top of the melt and may be separated by careful skimming. Moreover, this primary crystallization is subject to severe undercooling so that ordinary methods of thermal analysis are unsatisfactory for determining the liquidus. Dip sampling is applicable and yields results of great accuracy.

## Location of Boundaries of Liquid Immiscibility

Where reliable methods of chemical analysis are available, the location of the boundaries of two-liquid (or multiliquid) fields is a very simple matter. It is necessary only to hold the mixture for a long time at each of a series of fixed temperatures, drawing samples for analysis from each of the layers after equilibrium has been established at each temperature. This may be done first with the temperature advancing in each step and later with the temperature decreasing in order to obtain an approach to equilibrium from two directions.

If suitable analytical methods are lacking, it is still possible to obtain an accurate measure of the location of the boundaries of liquid immiscibility. The metals to be used are melted in a cylindrical crucible of smooth and uniform bore. Great care must be taken to measure accurately the quantities of the metals used and to prevent any loss or contamination through oxidation, vaporization, or the incorporation of fluxes. Equilibrium is established at a fixed temperature, and the crucible containing the two liquid layers is cooled quickly until its contents are solid. The composite ingot is removed from the crucible, separated at the interface, and each layer weighed. These weights will be proportional to the lengths of the arms of a tie-line lever with its fulcrum at the gross composition of the mixture $X$, Fig. 21-13, and its ends on the two boundaries of the two-liquid field $L_I$ and $L_{II}$. The absolute length of the tie-line $L_I L_{II}$ must be determined by a second experiment in which the proportions of the two metals are changed. A mixture of the same total weight of composition $Y$

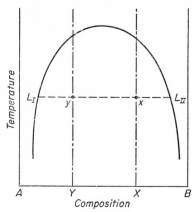

is brought to equilibrium at the same temperature, cooled, and sectioned, and the two layers weighed. Now the difference in weight between layer 1 in the two samples (or between layer 2 in the two samples) is proportional to the known difference in the gross compositions of the two samples, $XY$. This gives a ratio which, applied to the weight of each layer in either experiment, gives the actual lengths of the lever arms in weight percentage of the components and thereby the true compositions of the conjugate

FIG. 21-13. Illustrating a method for determining the position of the boundary of two-liquid immiscibility.

liquids. Repeated at a series of temperatures, this method is capable of giving an exact determination of the boundaries of the two-liquid region.

## The Fracture at Liquation Method

An extremely sensitive, and at the same time simple, method for locating solidus points and eutectic temperatures is by observing the temperature at which a sample, in the form of a bar, fractures under light static load during slow heating. Nothing at all elaborate in the way of apparatus is needed. Indeed, it is sometimes sufficient to support the sample in a muffle furnace on a brick with another brick on top so that one end of the bar is projecting. A small weight is laid upon the projecting end, and a thermocouple is placed upon the bar at a point near the bricks. The furnace is heated slowly. When the bar suddenly sags and breaks, the temperature is read and recorded as the solidus temperature. Better temperature control may be achieved by using a stirred oil or salt bath or circulating-air furnace. A jig to hold and load the specimen makes it easier to duplicate experimental conditions.[1] It is an easy matter also to make a continuous record of the temperature with an automatic recorder and so to arrange the apparatus that the thermocouple is short-circuited by the displacement of the breaking bar, thus making a positive indication on the recording chart. These refinements permit the use of very slow heating rates, which are highly desirable.

In order to avoid fracture at a temperature below the true solidus, it is

[1] See W. S. Pellini and F. N. Rhines, Constitution of Lead-rich Lead Antimony Alloys, *Trans. Am. Inst. Mining Met. Engrs.*, **152**:65–74 (1943).

advisable to homogenize the sample thoroughly before testing. This step may be omitted, however, if the heating rate is very slow. Occasionally, the test fails because the liquid phase does not wet the grain boundaries; duplicate samples then fracture at widely separated temperatures. When this happens, the method must be discarded. If all the samples tested (including at least three of each composition) fracture within less than a degree of the same temperature, and if the fracture is clearly intercrystalline, it is usually safe to accept the results.

## Dilatometry

Phase changes are usually accompanied by volume changes. If the length of a bar of an alloy is recorded continuously as a function of the temperature, inflections and arrests will be found in the temperature vs. length plot, Fig. 21-14, corresponding to phase boundaries and isothermal reactions in the solid state. Undercooling and overheating effects, being

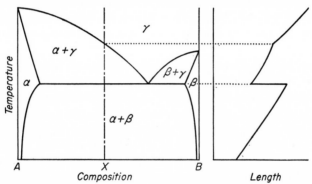

FIG. 21-14. Use of dilatometric measurements to determine points on phase boundaries in the solid state.

especially pronounced in transformations in the solid state, are usually observed to a marked degree in dilatometric studies. The characteristics of the method are such, however, that extremely slow heating and cooling rates may be used in order to minimize error from this source.

By the use of various kinds of lever systems the length changes can be magnified, for observation, to any desired degree. This gives to dilatometry a very high order of sensitivity which makes it possible to detect solid-state phase changes involving very small quantities of matter. The simplest dilatometer usually consists of a silica tube with one end closed and provision for the sample to rest firmly upon the closed end inside the tube. A silica rod rests upon the other end of the sample. (Silica is used for this purpose because of its very low coefficient of thermal expansion.) A dial gage secured to the outer tube and with its lever arm resting upon the protruding end of the silica rod serves to record length changes. That portion of the tube containing the sample may then be enclosed in a suitable furnace for temperature control. Greater sensitivity is obtained by substituting an optical lever for the dial gage.[1] In some cases interferometers have been employed for this purpose. These refinements are useful chiefly in differential dilatometry, wherein the length of the sample is compared with that of a bar of some metal of nearly the same expansivity but which undergoes no phase change within the temperature interval of interest.

## Metallographic Methods

There are many ways in which metallographic techniques can be applied to constitutional investigations. Regardless of the methods used in determining the phase diagram it is good practice, and usual, to verify the findings metallographically. The chief advantage of the metallographic method lies in its directness; we tend to place more confidence in that which we can see than in that which we infer from cooling curves, dilatometric curves, and the like. The chief disadvantages of the method are (1) that it is difficult to apply to metals while at high temperature and (2) that considerable skill and mature judgment are necessary for its reliable use. The difficulty of observing microstructures at high temperature detracts severely from the directness of the method, because "quenched samples" do not always retain their high-temperature structure.

One of the most common uses of the method is in the location of bivariant equilibria, illustrated in Fig. 21-15. Pieces of alloy $X$ are held for long periods, each at a different temperature, from $T_1$ to $T_9$. At the end of the heat treatments the samples are cooled quickly to retain their high-temperature structure and are examined metallographically. In the example of Fig. 21-15, those samples treated at $T_7$, $T_8$, and $T_9$ will exhibit two phases, $\alpha$ and $\beta$; those heated at $T_4$, $T_5$, and $T_6$ will be composed of $\alpha$ alone; while those heated at $T_1$, $T_2$, and $T_3$ will show structural evidence of liquation. From these observations it is concluded that the solvus, at composition $X$, lies between $T_6$ and $T_7$ and that the solidus lies between $T_3$ and $T_4$. If heat-treating temperatures that are very close together are used, the uncertainty in the transformation temperature may be reduced to very narrow limits. Complete solvus and solidus curves are obtained by repeating this process with a series of alloys of progressively changing composition. Good practice calls for the approach to equilibrium from two directions. For example, one set of samples that is to be used for locating the solvus may be homogenized at $T_5$ before being stabilized at

[1] See R. F. Mehl and C. Wells, Constitution of High Purity Iron-Carbon Alloys, *Trans. Am. Inst. Mining Met. Engrs.*, **125**:429–472. (1937).

the final temperature, while another set may be heat-treated below $T_9$ to produce a two-phase structure before the final heat treatment. Thus, in one set of samples, equilibrium is being approached by the precipitation of the $\beta$ phase, while in the other set, the $\beta$ phase is going into solid solution.

The metallographic method is well adapted to general survey work. One scheme, that has already been discussed, involves the use of diffusion couples (see Fig. 11-8 and Fig. 13-37). The number of phases that occurs across an entire binary diagram at a given temperature can be ascer-

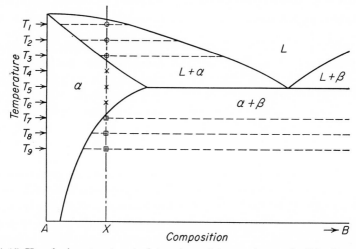

FIG. 21-15. Use of microstructure to determine points on solvus and solidus boundaries.

tained in this way. Similarly, the phases occurring in a ternary isotherm may, in principle, be determined by the use of a diffusion "triple," i.e., a diffusion sample made by bringing the three components together at a common point. Rather long diffusion times are necessary, and there is always uncertainty with regard to whether some very thin layer in the structure has remained undetected or two or more layers of similar appearance have not been differentiated. On the whole, the method is best suited for the *verification* of the succession of phases.

Another procedure consists in the examination of cast and heat-treated alloys at small composition intervals across the system. This is very laborious but usually worth the effort, not simply as a preliminary survey, but especially as a final check on the conclusions from other methods of investigation.

## X-ray Diffraction Methods

There are two essentially different ways in which X-ray diffraction is applied to constitutional studies. One is in the identification of crystalline phases, and the other is in the location of bivariant boundaries wholly within the solid state.

Each type of crystal gives a characteristic diffraction pattern when irradiated with a beam of X rays. The various types of crystals can be distinguished by the number and succession of their characteristic reflections. Crystals of the same type but differing in lattice dimensions can be distinguished by the spacing of their characteristic reflections. It is usually possible, therefore, to identify each solid phase in a polyphase alloy by comparing the diffraction pattern of the alloy with standard patterns that are known to represent individual phases. Tables giving the characteristic reflections of all the metals and many intermediate phases are available.[1] Limitations on the method are imposed by the superposition of reflections in polyphase alloys, sometimes making the identification of a phase uncertain, and by the rather large amount of a phase that must be present in order to detect its diffraction pattern. Less than 10% of a phase is difficult to identify, and less than 1% is hardly possible to detect.

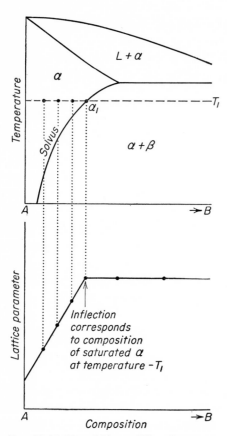

Fig. 21-16. Use of X-ray diffraction measurements, of the lattice parameter, to determine a point on a solvus curve.

Lattice parameter measurements are used for the location of solvus lines and other bivariant boundaries. The method rests upon the fact that the crystal lattice dimensions vary with composition in solid solutions. If the lattice parameter of an alloy of the terminal solid-solution type is measured by X-ray diffraction methods, it will be found that the parameter of the $\alpha$ phase changes progressively across the $\alpha$ field but remains constant in the $\alpha + \beta$ field (Fig. 21-16). The latter observation is to be expected because the composition of the $\alpha$ phase at any given tem-

[1] "X-ray Diffraction Data Card File," American Society for Testing Materials, Philadelphia, 1953.

perature in the two-phase field is fixed at the solvus boundary. If the lattice parameter of the $\alpha$ phase is first determined as a function of composition, then a single measurement of the parameter of the $\alpha$ phase in an alloy that has been stabilized within the $\alpha + \beta$ field will identify the composition of the $\alpha$ phase on the solvus line at the temperature of stabilization. By heat treating an $\alpha + \beta$ alloy at a series of temperatures and measuring the lattice parameter after each treatment, a succession of points on the solvus boundary is obtained. The method is both simple and capable of great precision in its application to binary phase diagrams.

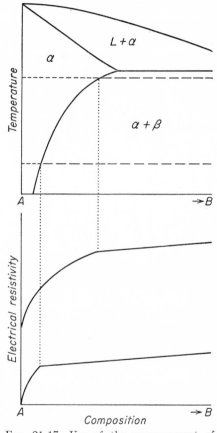

Nothing will be said here of the techniques of the X-ray diffraction methods themselves. This is a highly specialized field requiring extensive knowledge and skill, and there are excellent books devoted to the subject.[1]

## Electrical Resistivity Method

Another technique that is adapted to the location of bivariant boundaries and univariant isotherms in the solid state involves the measurement of electrical resistivity (or conductivity). Like the lattice parameter, electrical resistivity is a property that varies markedly with composition change in solid solutions (see Figs. 3-12 and 4-21). Unlike the lattice parameter, the measured resistivity in a polyphase mixture is not char-

Fig. 21-17. Use of the measurement of electrical resistivity, as a function of composition, to determine points on a solvus curve.

acteristic of any one of the conjugate phases but approximates the average resistivity of the mixture.

In application, the resistivity of a series of alloys is measured either at

[1] See, for example, C. S. Barrett, "Structure of Metals," McGraw-Hill Book Company, Inc., New York, 1952.

the temperature of investigation or after quenching from a long heat treatment at that temperature. It is preferable to make the measurements at temperature, and this can usually be done with little difficulty. The resistivity is then plotted as a function of composition for a fixed temperature (Fig. 21-17) or as a function of temperature at fixed composition (Fig. 21-18). In either case inflections in the curve mark the crossing of bivariant boundaries. The temperature-resistivity curve will show an "arrest" at each univariant transformation.

Relatively simple apparatus is used in this type of work. The sample should be in the form of wire, bar, or strip, well worked and annealed to

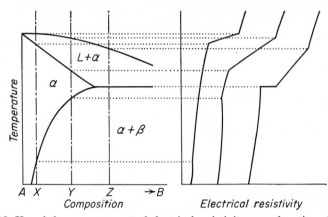

Fig. 21-18. Use of the measurement of electrical resistivity, as a function of temperature, to determine points on solvus, solidus, and lines of three-phase equilibrium.

minimize inhomogeneity. Leads for introducing the current are attached at the extreme ends, and potential leads are attached some distance inward from either end. Where measurements are to be made at elevated temperature, special precautions may be necessary to maintain good electrical contacts and to avoid contamination of the specimen by the lead wires. A good storage battery is needed to supply a steady current through the specimen and a potentiometer arranged to measure alternately the potential drop across a standard resistance in series with the specimen and the potential between the potential leads that are attached to the specimen. From the first reading the current is computed, and from this and the potential measured at the specimen the resistivity of the alloy is computed. For the measurement of true resistivity it is necessary to know accurately the length and cross section of the sample. For constitutional studies it is enough, however, to measure the resistivity relative to that of other samples in the series, and this requires only that the samples all be of like shape and size; it is not essential that their dimensions be known. Care should be taken to avoid heating the specimen by the use of too large a current applied for too long a time. The need for precautions to obtain equilibrium, and to make observations on approach to equilibrium from two directions, obtains here as elsewhere.

## Miscellaneous Methods

Almost any of the physical properties that are responsive to phase changes may be useful in the determination of phase diagrams. Measurements of mechanical properties have been employed at times for this purpose, but these are sensitive to so many factors in addition to phase changes that their use must remain very limited. Electrochemical methods have been developed[1] and have been found highly satisfactory for locating bivariant and tervariant boundaries in certain binary and ternary alloys. Very special skill and knowledge are required, however, to adapt the method to new systems.

[1] R. H. Brown, W. L. Fink, and M. S. Hunter, Measurement of Irreversible Potentials as a Metallurgical Research Tool, *Trans. Am. Inst. Mining Met. Engrs.*, **143**:115–123 (1941).

APPENDIX I

# GREEK ALPHABET

| | | | | |
|---|---|---|---|---|
| Alpha | $\alpha$ | Nu | $\nu$ |
| Beta | $\beta$ | Xi | $\xi$ |
| Gamma | $\gamma$ | Omicron | $o$ |
| Delta | $\delta$ | Pi | $\pi$ |
| Epsilon | $\epsilon$ | Rho | $\rho$ |
| Zeta | $\zeta$ | Sigma | $\sigma$ |
| Eta | $\eta$ | Tau | $\tau$ |
| Theta | $\theta$ | Upsilon | $\upsilon$ |
| Iota | $\iota$ | Phi | $\varphi$ |
| Kappa | $\kappa$ | Chi | $\chi$ |
| Lambda | $\lambda$ | Psi | $\psi$ |
| Mu | $\mu$ | Omega | $\omega$ |

# ATOMIC WEIGHTS OF THE ELEMENTS

| | | | |
|---|---|---|---|
| A...... 39.944 | Dy.... 162.46 | Mo.... 95.95 | S...... 32.066 |
| Ac..... 227.05 | Er.... 167.2 | N...... 14.008 | Sb.... 121.76 |
| Ag.... 107.880 | Eu.... 152.0 | Na.... 22.997 | Sc..... 45.10 |
| Al...... 26.97 | F...... 19.00 | Nd.... 144.27 | Se..... 78.96 |
| Am.... 241 | Fa.... 223 | Ne.... 20.183 | Si..... 28.06 |
| As..... 74.91 | Fe.... 55.85 | Ni.... 58.69 | Sm..... 150.43 |
| At..... 211 | Ga.... 69.72 | Np.... 237 | Sn.... 118.70 |
| Au..... 197.2 | Gc.... 156.9 | O...... 16.000 | Sr..... 87.63 |
| B...... 10.82 | Ge.... 72.60 | Os..... 190.2 | Ta.... 180.88 |
| Ba.... 137.36 | H...... 1.0080 | P...... 30.98 | Tb..... 159.2 |
| Be.... 9.02 | He.... 4.003 | Pa..... 231 | Tc..... 99 |
| Bi. .... 209.00 | Hf.... 178.6 | Pb.... 207.21 | Te.... 127.61 |
| Br.... 79.916 | Hg.... 200.61 | Pd.... 106.7 | Th.... 232.12 |
| C...... 12.010 | Ho.... 164.94 | Pm..... 147 | Ti...... 47.90 |
| Ca.... 40.08 | I...... 126.92 | Po.... 210 | Tl...... 204.39 |
| Cb.... 92.91 | In...... 114.76 | Pr...... 140.92 | Tm..... 169.4 |
| Cd.... 112.41 | Ir...... 193.1 | Pt.... 195.23 | U...... 238.07 |
| Ce.... 140.13 | K...... 39.096 | Pu.... 239 | V...... 50.95 |
| Cl...... 35.457 | Kr.... 83.7 | Ra.... 226.05 | W...... 183.92 |
| Cm.... 242 | La.... 138.92 | Rb.... 85.48 | Xe..... 131.3 |
| Co.... 58.94 | Li ... 6.940 | Re.... 186.31 | Y...... 88.92 |
| Cr.... 52.01 | Lu.... 174.99 | Rh.... 102.91 | Yb.... 173.04 |
| Cs.... 132.91 | Mg.... 24.32 | Rn.... 222 | Zn .... 65.38 |
| Cu.... 63.54 | Mn.... 54.93 | Ru.... 101.7 | Zr...... 91.22 |

# THE INTERCONVERSION OF ATOMIC, WEIGHT, AND VOLUME PERCENTAGES IN BINARY AND TERNARY SYSTEMS*

By Cyril Stanley Smith,† Waterbury, Conn.

In the study of the structure or the properties of a series of alloys or nonmetallic compounds, it is often advantageous to express the composition not as percentage by weight, according to which the material is always made up and analyzed, but as the atomic percentage of the two constituents. The conversion from the one to the other is represented by the formulas:

$$y = \frac{100 \, \frac{x}{A}}{\frac{x}{A} + \frac{100 - x}{B}} \text{ or } \frac{100x}{x + \frac{A}{B}(100 - x)} \qquad [1]$$

$$x = \frac{100yA}{yA + (100 - y)B} \text{ or } \frac{100y}{y + \frac{B}{A}(100 - y)} \qquad [2]$$

where $x$ and $y$ are the weight and atomic percentages respectively of the element of atomic weight $A$, and $(100 - x)$ and $(100 - y)$ are the percentages by weight and by atoms respectively of the element of atomic weight $B$.

While calculations according to these formulas are simple, they are time consuming if several conversions have to be made. In 1931 Waterman[1] constructed a nomograph for performing the conversion, but this did not permit an accuracy greater than about 1 per cent to be obtained in the neighborhood of 50 per cent. Ölander[2] has proposed a graphic method simpler than Waterman's and one that gives a uniform precision throughout the entire scale. The possible accuracy is about 0.5 per cent of the whole, depending on the size and accuracy of the coordinate paper used. Recently, Hermann[3]

---

* Note: Republished with the permission of the author and of the original publisher. This was originally published by the American Institute of Mining and Metallurgical Engineers, as Contribution No. 60, October 1933. Footnotes in this appendix are taken from the original article.

† Formerly: Copper Alloys Research Laboratory, The American Brass Co.; currently: Director, Institute for the Study of Metals, University of Chicago.

[1] Waterman: *Ind. & Eng. Chem.* (1931) **23**, 803.
[2] Ölander: *Ind. & Eng. Chem.*, Anal. Ed. (1932) **4**, 438.
[3] Herman: *Metallwirtschaft* (1933) **12**, 104.

has described a slide rule for making the calculations. These devices make use of the simple rearranged equation:

$$\frac{x}{100 - x} = \frac{A}{B}\left(\frac{y}{100 - y}\right) \tag{3}$$

The slide rule designed by Hermann has a stationary scale with the atomic weights of the elements on a logarithmic scale, the distances therefore being proportional to log $(A/B)$. The sliding scale is graduated with values of log $x/(100 - x)$ and is in two parts, corresponding to values of $x$ from 3 to 75 on one side and 0.3 to 23 on the reverse. This permits conversion from weight to atomic per cent, or vice versa, of any pair of elements with an accuracy of about 0.1 per cent of the total in the neighborhood of 10 per cent, decreasing to rather less than 0.3 per cent in the neighborhood of 50 per cent.

The function log $x/(100 - x)$ is equal to $-$log $[(100/x) - 1]$. This function is readily obtained by three operations involving the use of tables of reciprocals and ordinary logarithms, and provides a useful short cut in the calculations. It is possible to simplify the computation even further by the construction of a table of values of log $x/(100 - x)$ for values of $x$ in steps of 0.1 per cent from 0.1 to 99.9; interpolation permitting calculations accurate to 0.01 per cent of the whole in any part of the range. Such a table has been calculated by the author and is given in Table 1. To avoid negative characteristics, log $1000x/(100 - x)$ has been used [log $x/(100 - x) + 3$]. Great care was exercised in the computation of this table and it is believed to be correct. The values of $1000x/(100 - x)$ were originally computed by the use of five figure logarithms but they are given only to a sufficient number of places to permit interpolation accurate to 0.01 in the value of $x$. This required four places of decimals in most of the table, but rather less where the function changes rapidly. The difference figures are to be applied to the last two places of the figure next *below*—a point that should be borne in mind where the number of places of decimals in the main function changes.

Table 2 gives values of log $(A/B)$ for most pairs of elements likely to prove of importance in the study of alloy systems. It is obviously a simple matter to calculate such log ratios for any other systems of interest or to bring the present ones up to date if the atomic weight values are revised. The worker in a special field would be well advised to recalculate the values for his particular system, for it is unlikely that the table is entirely free from errors in spite of the care with which it was computed and checked. It is also possible to calculate ratios for determining the atomic composition of salts, using the radical as $A$, or the molecular concentrations in mixtures of salts or any compounds of known molecular weight.

The use of the tables is identical with the use of ordinary logarithmic tables for simple multiplication or division. The value of log $1000x/(100 - x)$ corresponding to the percentage, either weight or atomic, that it is desired to convert to the other is found from Table 1. To this figure is *added* the value of log $(A/B)$ (from Table 2) if the conversion is from *atomic* to weight per cent, or *subtracted* if the conversion is from *weight* to atomic per cent. The result is the value of log $1000x/(100 - x)$ corresponding to the percentage desired, which is then ascertained by referring to Table 1 again. In selecting the ratio $A/B$, $A$ is the element whose percentage amount is being converted, $B$ the other, usually the principal element. The reverse notation may be used if the log is subtracted instead of added, or vice versa.

In conversions of percentages less than 0.3 per cent it is better to use the simple ratio $x = (A/B)y$ because in that range the difference values become practically useless on account of the rapid decrease in log $1000x/(100 - x)$.

Two examples will be given to illustrate the use of the tables:

EXAMPLE I. An aluminum-zinc alloy contains 10.17 per cent zinc. What is the atomic percentage of zinc?

*Solution:* From Table 1, $\log \dfrac{1000x}{100 - x}$ for $x = 10.17$ is 2.0539

From Table 2, $\log \dfrac{\text{at. wt. Zn}}{\text{at. wt. Al}}$            is 0.3846

Subtracting, since we are converting from *weight* to *atomic* percentage, leaves                                1.6693

$$1.669(3) \text{ is } \log \frac{1000x}{100 - x} \text{ for } x = 4.44$$

∴ 10.17 wt. of Zn is 4.44 atomic per cent Zn in Al.

EXAMPLE II. What is the percentage of Si, by weight, in the compound $Cu_5Si$?

*Solution:* $Cu_5Si = 16.67$ atomic per cent Si

$$\log \frac{1000x}{100 - x} \text{ for } x = 16.67 = 2.3011$$

$$\log \frac{\text{at. wt. Si}}{\text{at. wt. Cu}} = \bar{1}.6448$$

Adding, since we are converting from *atomic* to *weight* per cent, gives                                1.9459

$$1.9459 = \log \frac{1000x}{100 - x} \text{ for } x = 8.11$$

∴ $Cu_5Si$ contains 8.11 per cent Si by weight.

## CONVERSION IN TERNARY SYSTEMS

Table I is not of much value for conversions involving three components, for it is necessary to compute an average atomic weight for two elements for every composition used and this is as tedious as the complete calculation by ordinary methods.

In cases where high accuracy is not required, as, for example, in the transposition of ternary equilibrium diagrams, graphical methods may be used to advantage. Hume-Rothery[1] has proposed an ingenious method involving the use of proportional compasses and a parallel ruler that is very suitable for this purpose. By a combination of the methods of Ölander[2] and Waterman[3] the present author has devised an alternative method which requires the use of no special instruments.

Ölander utilized the intercept theorem of Menelaos, the application of which to graphical conversions in binary systems is obvious from Fig. 1. Fig. 2 shows that a line, $Ae$, from the apex of the conventional equilateral triangle representing composition in a ternary system is the locus of the composition of all alloys with a given ratio, $Be/eC$, of two of the three components. The point $p$ represents a particular composition, and the line $Ape$ all alloys of the same proportion of the two components $B$ and $C$. Returning to Fig. 1, the point $c$ is arbitrarily selected, and for our present purposes we may place it so that the angle $obc$ is 60° and the line $cb$ equal to the side of the equilateral triangle used to represent the ternary system. By using a suitable atomic weight scale so that $ob$ equals $bc$, $ob$ becomes a second side of the equilateral triangle. When this is done, the line $oe$ of Fig. 1 becomes the line $Ae$ of Fig. 2, and serves the

[1] Hume-Rothery: *Paper* No. 643, Inst. of Metals, London (September, 1933).
[2,3] *Loc. cit.*

double purpose of projecting the composition of $p$ on to the side $BC$ of the triangle (Fig. 2), i. e., finding the binary ratio by weight of components $B$ and $C$ in the alloy, and, by intersection with the line $ca$ (Fig. 1), simultaneously determining the atomic ratio of $B$ and $C$ in the alloy. If this latter ratio is transferred to the side of the equilateral triangle by a line parallel to $ob$ and this point then joined to the opposite apex, the resulting line represents all alloys with the correct atomic binary ratio of $B$ and $C$. A similar operation with another pair of components gives a second binary atomic

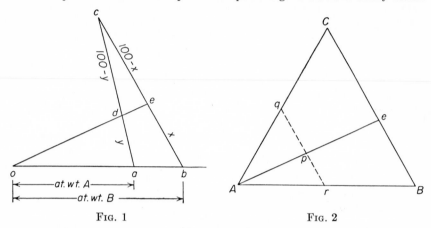

FIG. 1　　　　　　　　　　FIG. 2

ratio line and the point of intersection of the two obviously corresponds to the composition of the alloy, in atomic percentages of the three elements.

$$\frac{be}{ec} = \frac{ad}{dc} \times \frac{ob}{oa}$$

$$\frac{x}{100 - x} = \frac{y}{100 - y} \times \frac{\text{at. wt. } B}{\text{at. wt. } A}$$

($x$ and $y$ are weight and atomic percentages of component $\overline{B}$)

$$\text{FIG. 2.—} \frac{rp}{pq} = \frac{Be}{eC}$$

The method is demonstrated in Fig. 3. For conversion on the side $BC$, i. e., utilizing the ratio of components $B$ and $C$, draw the line $CB'$, $B'$ dividing the length $AB$ (now the atomic weight scale for $B$ and $C$) in the proportion of the atomic weights of components $B$ and $C$,[1] so that $AB' = AB \times$ (at. wt. $B$)/(at. wt. $C$). Through any point $p$ that is to be converted from weight percentage to atomic percentage draw the line $Apde$. $e$ then corresponds to the weight ratio of the two components $B$ and $C$, $d$ to the atomic ratio. Draw $de'$, parallel to $AB$, then $Ae'$, which represents all alloys of the same atomic ratio of components $B$ and $C$ as the alloy of weight percentage shown by $p$.

[1] The point $B'$ is not necessarily between $A$ and $B$, but it is simpler to maintain all operations within the triangle, and for this reason the atomic weight scale should be set up on the side of the element of lower atomic weight. Note that the point $B$ on the side $AB$ when this is considered as an atomic weight scale corresponds to the atomic weight of element $C$, and the atomic weight of $B$ is represented as $B'$. $A$ corresponds to zero atomic weight for conversions of ratios of $B$ and $C$, but becomes atomic weight $B$ on the scale $CA$ when dealing with the pair of elements $B$ and $A$. Fig. 1 should be kept in mind, for the weight percentages of an element are represented on the line joining the point $c$ to the atomic weight of the element concerned, and the atomic percentage on the line joining $c$ to the other atomic weight.

Choosing a second pair of components, for example, $A$ and $B$, draw the line $BA'$, $A'$ being located so that $CA' = CA \times$ (at. wt. $A$)/(at. wt. $B$). The intersection of $Cp$ with $BA'$ gives $f$, transferred to the side $AB$ as $g'$. $Cg'$ represents all compositions of the same atomic ratio of $A$ and $B$ as the alloy $p$, and where this line intersects the other atomic ratio line, $Ae'$, lies the point $p'$, corresponding to the atomic composition of the alloy whose composition by weight is given by $p$.

In practice, when using triangular coordinate paper with fairly close rulings, the lines $Apde$ and $de'$ will not be drawn, but the point $d$ obtained by a straight edge and transferred to $e'$ by eye. Only part of $Ae'$ and $Cg'$ will be drawn, enough to establish the intersection. The time involved in setting off the points $B'$ and $A'$ is short, and once the lines $CB'$ and $BA'$ have been drawn they are used for any number of conversions in the system.

Conversion from atomic to weight percentages can be made with equal facility,

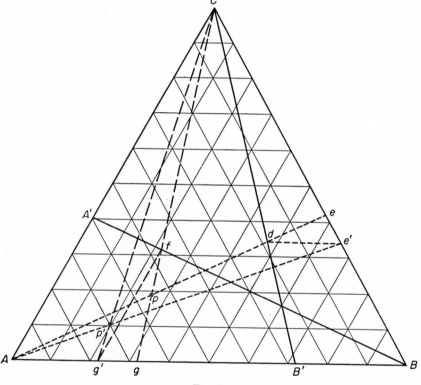

FIG. 3

either by working in the reverse direction from the example given above, or by setting up the points $B'$ and $A'$ so that $AB' = AB$ (at. wt. $C$)/(at. wt. $B$) and

$$CA' = CA \times \frac{\text{at. wt. } B}{\text{at. wt. } A}$$

respectively, and proceeding as in Fig. 3. Weight percentages may be converted to volume percentages by the same methods, if the points $B'$ and $A'$ are located so that

$$AB' = AB \times \frac{\text{density } B}{\text{density } C} \qquad \text{and} \qquad CA' = CA \times \frac{\text{density } A}{\text{density } B}$$

PHASE DIAGRAMS IN METALLURGY

Table 1. Values of Log $\dfrac{x}{100-x}+3$

| × | 0 | 0.1 | 0.2 | 0.3 | 0.4 | 0.5 | 0.6 | 0.7 | 0.8 | 0.9 |
|---|---|---|---|---|---|---|---|---|---|---|
| 0 | ∞ | 0.00 *30* | 30 *18* | 48 *12* | 604 *97* | 701 *80* | 781 *67* | 848 *58* | 906 *50* | 958 *48* |
| 1 | 1.004 *42* | 046 *38* | 084 *36* | 120 *32* | 152 *31* | 183 *28* | 211 *27* | 238 *25* | 263 *24* | 287 *23* |
| 2 | 310 *21* | 331 *21* | 352 *20* | 372 *19* | 391 *18* | 409 *17* | 426 *17* | 443 *16* | 459 *16* | 475 *15* |
| 3 | 490 *15* | 505 *14* | 519 *14* | 533 *14* | 547 *13* | 560 *13* | 572 *12* | 585 *12* | 597 *11* | 608 *12* |
| 4 | 620 *11* | 631 *11* | 642 *11* | 653 *10* | 663 *10* | 673 *10* | 6832 *98* | 6930 *96* | 7026 *94* | 7120 *92* |
| 5 | 1.7212 *91* | 7303 *89* | 7392 *87* | 7479 *86* | 7565 *84* | 7649 *83* | 7732 *82* | 7814 *80* | 7894 *79* | 7973 *77* |
| 6 | 8050 *77* | 8127 *75* | 8202 *74* | 8276 *73* | 8349 *72* | 8421 *71* | 8492 *70* | 8562 *69* | 8631 *68* | 8699 *67* |
| 7 | 8766 *66* | 8832 *65* | 8898 *64* | 8962 *64* | 9026 *63* | 9089 *62* | 9151 *62* | 9213 *61* | 9274 *60* | 9334 *59* |
| 8 | 9393 *59* | 9452 *58* | 9510 *57* | 9567 *57* | 9624 *56* | 9680 *56* | 9736 *55* | 9791 *54* | 9845 *54* | 9899 *53* |
| 9 | 9952 *53* | **0005** *52* | **0057** *52* | **0109** *51* | **0160** *51* | **0211** *50* | **0261** *50* | **0311** *49* | **0360** *49* | **0409** *49* |
| 10 | 2.0458 *48* | 0506 *47* | 0553 *47* | 0600 *47* | 0647 *47* | 0694 *46* | 0740 *45* | 0785 *46* | 0831 *45* | 0876 *44* |
| 11 | 0920 *44* | 0964 *44* | 1008 *44* | 1052 *43* | 1095 *43* | 1138 *42* | 1180 *42* | 1222 *42* | 1264 *42* | 1306 *41* |
| 12 | 1347 *41* | 1388 *41* | 1429 *40* | 1469 *40* | 1509 *40* | 1549 *40* | 1589 *39* | 1628 *39* | 1667 *39* | 1706 *38* |
| 13 | 1744 *39* | 1783 *38* | 1821 *37* | 1858 *38* | 1896 *37* | 1933 *37* | 1970 *37* | 2007 *37* | 2044 *36* | 2080 *36* |
| 14 | 2116 *36* | 2152 *36* | 2188 *36* | 2224 *35* | 2259 *35* | 2294 *35* | 2329 *35* | 2364 *34* | 2398 *35* | 2433 *34* |
| 15 | 2.2467 *34* | 2501 *33* | 2534 *34* | 2568 *34* | 2602 *33* | 2635 *33* | 2668 *33* | 2701 *33* | 2734 *32* | 2766 *32* |
| 16 | 2798 *33* | 2831 *32* | 2863 *32* | 2895 *32* | 2926 *32* | 2958 *31* | 2989 *32* | 3021 *31* | 3052 *31* | 3083 *31* |
| 17 | 3114 *31* | 3145 *30* | 3175 *30* | 3205 *31* | 3236 *30* | 3266 *30* | 3296 *30* | 3326 *30* | 3356 *29* | 3385 *30* |
| 18 | 3415 *29* | 3444 *29* | 3473 *29* | 3502 *29* | 3531 *29* | 3560 *29* | 3589 *29* | 3618 *28* | 3646 *28* | 3674 *29* |
| 19 | 3703 *28* | 3731 *28* | 3759 *28* | 3787 *28* | 3815 *28* | 3842 *27* | 3870 *28* | 3898 *27* | 3925 *27* | 3952 *27* |
| 20 | 2.3979 *28* | 4007 *27* | 4034 *26* | 4060 *27* | 4087 *27* | 4114 *27* | 4141 *26* | 4167 *26* | 4193 *27* | 4220 *26* |
| 21 | 4246 *26* | 4272 *26* | 4298 *26* | 4324 *26* | 4350 *26* | 4376 *25* | 4401 *26* | 4427 *26* | 4453 *25* | 4478 *25* |
| 22 | 4503 *26* | 4529 *25* | 4554 *25* | 4579 *25* | 4604 *25* | 4629 *25* | 4654 *25* | 4679 *24* | 4703 *25* | 4728 *24* |
| 23 | 4752 *25* | 4777 *24* | 4801 *25* | 4826 *24* | 4850 *24* | 4874 *24* | 4898 *24* | 4922 *24* | 4946 *24* | 4970 *24* |
| 24 | 4994 *24* | 5018 *24* | 5042 *23* | 5065 *24* | 5089 *23* | 5112 *24* | 5136 *23* | 5159 *23* | 5182 *24* | 5206 *23* |
| 25 | 2.5229 *23* | 5252 *23* | 5275 *23* | 5298 *23* | 5321 *23* | 5344 *23* | 5367 *22* | 5389 *23* | 5412 *23* | 5435 *22* |
| 26 | 5457 *23* | 5480 *22* | 5502 *23* | 5525 *22* | 5547 *23* | 5570 *22* | 5592 *22* | 5614 *22* | 5636 *22* | 5658 *22* |
| 27 | 5680 *22* | 5702 *22* | 5724 *22* | 5746 *22* | 5768 *22* | 5790 *22* | 5812 *21* | 5833 *22* | 5855 *22* | 5877 *21* |
| 28 | 5898 *22* | 5920 *21* | 5941 *22* | 5963 *21* | 5984 *21* | 6005 *22* | 6027 *21* | 6048 *21* | 6069 *21* | 6090 *21* |
| 29 | 6111 *21* | 6132 *22* | 6154 *21* | 6175 *21* | 6196 *20* | 6216 *21* | 6237 *21* | 6258 *21* | 6279 *21* | 6300 *20* |
| 30 | 2.6320 *21* | 6341 *21* | 6362 *20* | 6382 *21* | 6403 *20* | 6423 *21* | 6444 *20* | 6464 *20* | 6484 *21* | 6505 *20* |
| 31 | 6525 *20* | 6545 *21* | 6566 *20* | 6586 *20* | 6606 *20* | 6626 *20* | 6646 *20* | 6666 *20* | 6686 *20* | 6706 *20* |
| 32 | 6726 *20* | 6746 *20* | 6766 *20* | 6786 *20* | 6806 *20* | 6826 *20* | 6846 *19* | 6865 *20* | 6885 *20* | 6905 *19* |
| 33 | 6924 *20* | 6944 *20* | 6964 *19* | 6983 *20* | 7003 *19* | 7022 *20* | 7042 *19* | 7061 *20* | 7081 *19* | 7100 *19* |
| 34 | 7119 *20* | 7139 *19* | 7158 *19* | 7177 *20* | 7197 *19* | 7216 *19* | 7235 *19* | 7254 *19* | 7273 *19* | 7293 *19* |
| 35 | 2.7312 *19* | 7331 *19* | 7350 *19* | 7369 *19* | 7388 *19* | 7407 *19* | 7426 *19* | 7445 *18* | 7463 *19* | 7482 *19* |
| 36 | 7501 *19* | 7520 *19* | 7539 *19* | 7558 *18* | 7576 *19* | 7595 *19* | 7614 *19* | 7633 *18* | 7651 *19* | 7670 *19* |
| 37 | 7689 *18* | 7707 *19* | 7726 *18* | 7744 *19* | 7763 *18* | 7781 *19* | 7800 *19* | 7819 *18* | 7837 *19* | 7856 *18* |
| 38 | 7874 *18* | 7892 *19* | 7911 *18* | 7929 *19* | 7948 *18* | 7966 *18* | 7984 *19* | 8003 *18* | 8021 *18* | 8039 *18* |
| 39 | 8057 *19* | 8076 *18* | 8094 *18* | 8112 *18* | 8130 *18* | 8148 *19* | 8167 *18* | 8185 *18* | 8203 *18* | 8221 *18* |
| 40 | 2.8239 *18* | 8257 *18* | 8275 *18* | 8293 *18* | 8311 *18* | 8329 *18* | 8347 *18* | 8365 *18* | 8383 *18* | 8401 *18* |
| 41 | 8419 *18* | 8437 *18* | 8455 *18* | 8473 *18* | 8491 *18* | 8509 *18* | 8527 *18* | 8545 *18* | 8563 *17* | 8580 *18* |
| 42 | 8598 *18* | 8616 *18* | 8634 *18* | 8652 *18* | 8670 *17* | 8687 *18* | 8705 *18* | 8723 *18* | 8741 *17* | 8758 *18* |
| 43 | 8776 *18* | 8794 *17* | 8811 *18* | 8829 *18* | 8847 *17* | 8864 *18* | 8882 *17* | 8900 *17* | 8917 *18* | 8935 *18* |
| 44 | 8953 *17* | 8970 *18* | 8988 *17* | 9005 *18* | 9023 *18* | 9041 *17* | 9058 *18* | 9076 *17* | 9093 *18* | 9111 *18* |
| 45 | 2.9129 *17* | 9146 *18* | 9164 *17* | 9181 *18* | 9199 *17* | 9216 *18* | 9234 *17* | 9251 *18* | 9269 *17* | 9286 *18* |
| 46 | 9304 *17* | 9321 *18* | 9339 *17* | 9356 *18* | 9374 *17* | 9391 *18* | 9409 *17* | 9426 *17* | 9443 *18* | 9461 *17* |
| 47 | 9478 *18* | 9496 *17* | 9513 *18* | 9531 *17* | 9548 *17* | 9565 *18* | 9583 *17* | 9600 *18* | 9618 *17* | 9635 *17* |
| 48 | 9652 *18* | 9670 *17* | 9687 *18* | 9705 *17* | 9722 *17* | 9739 *18* | 9757 *17* | 9774 *18* | 9792 *17* | 9809 *17* |
| 49 | 9826 *18* | 9844 *17* | 9861 *17* | 9878 *18* | 9896 *17* | 9913 *18* | 9931 *17* | 9948 *17* | 9965 *18* | 9983 *17* |

Subtract value of log $(A/B)$ (from table 2) if converting from weight to atomic percentage.

TABLE 1. VALUES OF LOG $\dfrac{x}{100-x} + 3$ (*Continued*)

| × | 0 | | 0.1 | | 0.2 | | 0.3 | | 0.4 | | 0.5 | | 0.6 | | 0.7 | | 0.8 | | 0.9 | |
|----|------|----|------|----|------|----|------|----|------|----|------|----|------|----|------|----|------|----|------|----|
| 50 | 3.0000 | *17* | 0017 | *18* | 0035 | *17* | 0052 | *18* | 0070 | *17* | 0087 | *17* | 0104 | *18* | 0122 | *17* | 0139 | *17* | 0156 | *18* |
| 51 | 0174 | *17* | 0191 | *18* | 0209 | *17* | 0226 | *17* | 0243 | *18* | 0261 | *17* | 0278 | *17* | 0295 | *18* | 0313 | *17* | 0330 | *18* |
| 52 | 0348 | *17* | 0365 | *17* | 0382 | *18* | 0400 | *17* | 0417 | *18* | 0435 | *17* | 0452 | *18* | 0470 | *17* | 0487 | *17* | 0504 | *18* |
| 53 | 0522 | *17* | 0539 | *18* | 0557 | *17* | 0574 | *18* | 0592 | *17* | 0609 | *17* | 0626 | *18* | 0644 | *17* | 0661 | *18* | 0679 | *17* |
| 54 | 0696 | *18* | 0714 | *17* | 0731 | *18* | 0749 | *17* | 0766 | *18* | 0784 | *17* | 0801 | *18* | 0819 | *17* | 0836 | *18* | 0854 | *18* |
| 55 | 3.0872 | *17* | 0889 | *18* | 0907 | *17* | 0924 | *18* | 0942 | *17* | 0959 | *18* | 0977 | *18* | 0995 | *17* | 1012 | *18* | 1030 | *17* |
| 56 | 1047 | *18* | 1065 | *18* | 1083 | *17* | 1100 | *18* | 1118 | *18* | 1136 | *17* | 1153 | *18* | 1171 | *18* | 1189 | *17* | 1206 | *18* |
| 57 | 1224 | *18* | 1242 | *18* | 1260 | *17* | 1277 | *18* | 1295 | *18* | 1313 | *18* | 1331 | *17* | 1348 | *18* | 1366 | *18* | 1384 | *18* |
| 58 | 1402 | *18* | 1420 | *17* | 1437 | *18* | 1455 | *18* | 1473 | *18* | 1491 | *18* | 1509 | *18* | 1527 | *18* | 1545 | *18* | 1563 | *18* |
| 59 | 1581 | *18* | 1599 | *18* | 1617 | *18* | 1635 | *18* | 1653 | *18* | 1671 | *18* | 1689 | *18* | 1707 | *18* | 1725 | *18* | 1743 | *18* |
| 60 | 3.1761 | *18* | 1779 | *18* | 1797 | *18* | 1815 | *18* | 1833 | *19* | 1852 | *18* | 1870 | *18* | 1888 | *18* | 1906 | *18* | 1924 | *19* |
| 61 | 1943 | *18* | 1961 | *18* | 1979 | *19* | 1998 | *18* | 2016 | *18* | 2034 | *19* | 2053 | *18* | 2071 | *18* | 2089 | *19* | 2108 | *18* |
| 62 | 2126 | *19* | 2145 | *18* | 2163 | *19* | 2182 | *18* | 2200 | *19* | 2219 | *18* | 2237 | *19* | 2256 | *18* | 2274 | *19* | 2293 | *18* |
| 63 | 2311 | *19* | 2330 | *19* | 2349 | *18* | 2367 | *19* | 2386 | *19* | 2405 | *19* | 2424 | *18* | 2442 | *19* | 2461 | *19* | 2480 | *19* |
| 64 | 2499 | *19* | 2518 | *19* | 2537 | *18* | 2555 | *19* | 2574 | *19* | 2593 | *19* | 2612 | *19* | 2631 | *19* | 2650 | *19* | 2669 | *19* |
| 65 | 3.2688 | *20* | 2708 | *19* | 2727 | *19* | 2746 | *19* | 2765 | *19* | 2784 | *19* | 2803 | *20* | 2823 | *19* | 2842 | *19* | 2861 | *20* |
| 66 | 2881 | *19* | 2900 | *20* | 2919 | *20* | 2939 | *19* | 2958 | *20* | 2978 | *19* | 2997 | *20* | 3017 | *19* | 3036 | *20* | 3056 | *20* |
| 67 | 3076 | *19* | 3095 | *20* | 3115 | *20* | 3135 | *19* | 3154 | *20* | 3174 | *20* | 3194 | *20* | 3214 | *20* | 3234 | *20* | 3254 | *20* |
| 68 | 3274 | *20* | 3294 | *20* | 3314 | *20* | 3334 | *20* | 3354 | *20* | 3374 | *20* | 3394 | *20* | 3414 | *20* | 3434 | *21* | 3455 | *20* |
| 69 | 3475 | *20* | 3495 | *21* | 3516 | *20* | 3536 | *20* | 3556 | *21* | 3577 | *20* | 3597 | *21* | 3618 | *21* | 3639 | *20* | 3659 | *21* |
| 70 | 3.3680 | *21* | 3701 | *20* | 3721 | *21* | 3742 | *21* | 3763 | *21* | 3784 | *21* | 3805 | *21* | 3826 | *21* | 3847 | *21* | 3868 | *21* |
| 71 | 3889 | *21* | 3910 | *21* | 3931 | *21* | 3952 | *21* | 3973 | *22* | 3995 | *21* | 4016 | *21* | 4037 | *22* | 4059 | *21* | 4080 | *22* |
| 72 | 4102 | *21* | 4123 | *22* | 4145 | *22* | 4167 | *21* | 4188 | *22* | 4210 | *22* | 4232 | *22* | 4254 | *22* | 4276 | *22* | 4298 | *22* |
| 73 | 4320 | *22* | 4342 | *22* | 4364 | *22* | 4386 | *22* | 4408 | *22* | 4430 | *23* | 4453 | *22* | 4475 | *22* | 4498 | *22* | 4520 | *23* |
| 74 | 4543 | *22* | 4565 | *23* | 4588 | *23* | 4611 | *22* | 4633 | *23* | 4656 | *23* | 4679 | *23* | 4702 | *23* | 4725 | *23* | 4748 | *23* |
| 75 | 3.4771 | *23* | 4794 | *24* | 4818 | *23* | 4841 | *23* | 4864 | *24* | 4888 | *23* | 4911 | *24* | 4935 | *24* | 4959 | *23* | 4982 | *24* |
| 76 | 5006 | *24* | 5030 | *24* | 5054 | *24* | 5078 | *24* | 5102 | *24* | 5126 | *24* | 5150 | *24* | 5174 | *25* | 5199 | *24* | 5223 | *25* |
| 77 | 5248 | *24* | 5272 | *25* | 5297 | *25* | 5322 | *24* | 5346 | *25* | 5371 | *25* | 5396 | *25* | 5421 | *25* | 5446 | *26* | 5472 | *25* |
| 78 | 5497 | *25* | 5522 | *26* | 5548 | *25* | 5573 | *26* | 5599 | *25* | 5624 | *26* | 5650 | *26* | 5676 | *26* | 5702 | *26* | 5728 | *26* |
| 79 | 5754 | *26* | 5780 | *27* | 5807 | *26* | 5833 | *27* | 5860 | *26* | 5886 | *27* | 5913 | *27* | 5940 | *27* | 5967 | *27* | 5994 | *27* |
| 80 | 3.6021 | *27* | 6048 | *27* | 6075 | *28* | 6103 | *27* | 6130 | *28* | 6158 | *27* | 6185 | *28* | 6213 | *28* | 6241 | *28* | 6269 | *28* |
| 81 | 6297 | *29* | 6326 | *28* | 6354 | *29* | 6383 | *28* | 6411 | *29* | 6440 | *29* | 6469 | *29* | 6498 | *29* | 6527 | *29* | 6556 | *29* |
| 82 | 6585 | *30* | 6615 | *30* | 6645 | *29* | 6674 | *30* | 6704 | *30* | 6734 | *30* | 6764 | *31* | 6795 | *30* | 6825 | *31* | 6856 | *30* |
| 83 | 6886 | *31* | 6917 | *31* | 6948 | *31* | 6979 | *32* | 7011 | *31* | 7042 | *32* | 7074 | *31* | 7105 | *32* | 7137 | *32* | 7169 | *33* |
| 84 | 7202 | *32* | 7234 | *33* | 7267 | *32* | 7299 | *33* | 7332 | *33* | 7365 | *33* | 7398 | *34* | 7432 | *34* | 7466 | *33* | 7499 | *34* |
| 85 | 3.7533 | *34* | 7567 | *35* | 7602 | *34* | 7636 | *35* | 7671 | *35* | 7706 | *35* | 7741 | *35* | 7776 | *36* | 7812 | *36* | 7848 | *36* |
| 86 | 7884 | *36* | 7920 | *36* | 7956 | *37* | 7993 | *37* | 8030 | *37* | 8067 | *37* | 8104 | *38* | 8142 | *38* | 8180 | *38* | 8218 | *38* |
| 87 | 8256 | *38* | 8294 | *39* | 8333 | *39* | 8372 | *39* | 8411 | *40* | 8451 | *40* | 8491 | *40* | 8531 | *40* | 8571 | *41* | 8612 | *41* |
| 88 | 8653 | *41* | 8694 | *42* | 8736 | *42* | 8778 | *42* | 8820 | *42* | 8862 | *43* | 8905 | *43* | 8948 | *44* | 8992 | *44* | 9036 | *44* |
| 89 | 9080 | *44* | 9124 | *45* | 9169 | *46* | 9215 | *45* | 9260 | *46* | 9306 | *47* | 9353 | *47* | 9400 | *47* | 9447 | *47* | 9494 | *48* |
| 90 | 3.9542 | *49* | 9591 | *49* | 9640 | *49* | 9689 | *50* | 9739 | *50* | 9789 | *51* | 9840 | *51* | 9891 | *52* | 9943 | *52* | 9995 | *53* |
| 91 | 4.0048 | *53* | 0101 | *54* | 0155 | *55* | 0210 | *55* | 0265 | *55* | 0320 | *56* | 0376 | *57* | 0433 | *57* | 0490 | *58* | 0548 | *59* |
| 92 | 0607 | *59* | 0666 | *60* | 0726 | *61* | 0787 | *62* | 0849 | *62* | 0911 | *63* | 0974 | *64* | 1038 | *64* | 1102 | *66* | 1168 | *66* |
| 93 | 1234 | *67* | 1301 | *68* | 1369 | *69* | 1438 | *70* | 1508 | *71* | 1579 | *72* | 1651 | *73* | 1724 | *74* | 1798 | *75* | 1873 | *77* |
| 94 | 1950 | *77* | 2027 | *79* | 2106 | *80* | 2186 | *82* | 2268 | *83* | 2351 | *84* | 2435 | *86* | 2521 | *87* | 2608 | *89* | 2697 | *91* |
| 95 | 4.2788 | *92* | 2880 | *94* | 2974 | *96* | 3070 | *98* | 317 | *10* | 327 | *10* | 337 | *10* | 347 | *11* | 358 | *11* | 369 | *11* |
| 96 | 380 | *12* | 392 | *11* | 403 | *12* | 415 | *13* | 428 | *12* | 440 | *13* | 453 | *14* | 467 | *14* | 481 | *14* | 495 | *15* |
| 97 | 510 | *15* | 525 | *16* | 541 | *16* | 557 | *17* | 574 | *17* | 591 | *18* | 609 | *19* | 628 | *20* | 648 | *21* | 669 | *21* |
| 98 | 690 | *23* | 713 | *24* | 737 | *25* | 762 | *27* | 789 | *28* | 817 | *31* | 848 | *32* | 880 | *36* | 916 | *38* | 954 | *42* |
| 99 | 996 | *46* | **042** | *51* | **093** | *59* | **152** | *67* | **219** | *79* | 299 | *97* | 40 | *12* | **52** | *18* | **70** | *30* | 6.00 | |

Add value of log $A/B$ when converting from atomic to weight percentage.

CONVERSIONS INVOLVING VOLUME PERCENTAGE

Table 1 may be used with a suitable factor to convert from either atomic or weight into volume percentages in any binary system. This depends upon the relation $x/(100 - x) =$ (density $A$/(density $B$) $[z(100 - z)]$ where $z$ is the volume percentage of the element $A$. Obviously, if we know the value for a given system of log [(density $A$/ (density $B$)] may be used in an identical manner to log $(A/B)$ for the atomic conversion and will give volume percentages directly. If the conversion is from atomic percentage to volume percentage, the value of log [(density $A$)/(density $B$) $\times$ (at. wt. $B$/ (at. wt. $A$)] is used. The log density function is *added* if the conversion is *from* volume percentage; subtracted if the conversion is *to* volume percentage from either weight or atomic percentage.

EXAMPLE III. What is the volume percentage of copper in a copper-iron alloy containing 4.82 per cent copper by weight?

*Solution:* From Table 1, $\log \dfrac{1000x}{100 - x}$ for $x = 4.82 = 0.7045$

$\quad$ Log $\dfrac{\text{density Cu}}{\text{density Fe}} = \log \left(\dfrac{8.94}{7.86}\right) \qquad = \qquad \underline{0.0559}$

$\quad$ Subtracting $\qquad\qquad\qquad\qquad\qquad = \qquad 0.6486$

$\qquad$ From Table 2, $0.649 = \log \dfrac{1000x}{100 - x}$ for $x = 4.26$

4.82 weight per cent of copper = 4.26 volume per cent copper in copper-iron alloy.

In view of the uncertainty of the available density figures for the elements, it was thought advisable not to compute a table of the log density ratios, but to leave the calculation of these to the workers in individual systems.

## Table 2. Values of Log Atomic Weight Ratios
### Element B

| At. No. | Symbol | 11 | 12 | 13 | 24 | 25 | 26 | 27 | 28 | 29 | 30 | 47 |
|---|---|---|---|---|---|---|---|---|---|---|---|---|
| | Symbol | Na | Mg | Al | Cr | Mn | Fe | Co | Ni | Cu | Zn | Ag |
| | At. Wt | 22.997 | 24.32 | 26.97 | 52.01 | 54.93 | 55.85 | 58.94 | 58.69 | 63.54 | 65.38 | 107.88 |
| | Log At. Wt | 1.3617 | 1.3860 | 1.4309 | 1.7161 | 1.7398 | 1.7470 | 1.7704 | 1.7686 | 1.8030 | 1.8154 | 2.0329 |

Element A:

| At. No. | Symbol | 11 Na | 12 Mg | 13 Al | 24 Cr | 25 Mn | 26 Fe | 27 Co | 28 Ni | 29 Cu | 30 Zn | 47 Ag |
|---|---|---|---|---|---|---|---|---|---|---|---|---|
| 1 | H | $\bar{2}$.6417 | $\bar{2}$.6174 | $\bar{2}$.5725 | $\bar{2}$.2873 | $\bar{2}$.2636 | $\bar{2}$.2564 | $\bar{2}$.2330 | $\bar{2}$.2348 | $\bar{2}$.2004 | $\bar{2}$.1888 | $\bar{3}$.9704 |
| 3 | Li | $\bar{1}$.4797 | $\bar{1}$.4554 | $\bar{1}$.4105 | $\bar{1}$.1353 | $\bar{1}$.1016 | $\bar{1}$.0944 | $\bar{1}$.0710 | $\bar{1}$.0728 | $\bar{1}$.0383 | $\bar{1}$.0259 | $\bar{2}$.8984 |
| 4 | Be | 5935 | 5693 | 5243 | 2391 | 2154 | 2083 | 1848 | 1867 | 1522 | 1398 | 9223 |
| 5 | B | $\bar{1}$.6726 | $\bar{1}$.6483 | $\bar{1}$.6034 | $\bar{1}$.3181 | $\bar{1}$.2944 | $\bar{1}$.2873 | $\bar{1}$.2638 | $\bar{1}$.2657 | $\bar{1}$.2312 | $\bar{1}$.2188 | $\bar{1}$.0013 |
| 6 | C | 7178 | 6935 | 6486 | 3634 | 3397 | 3325 | 3091 | 3109 | 2752 | 2644 | 0466 |
| 7 | N | 7847 | 7604 | 7155 | 4303 | 4066 | 3994 | 3760 | 3778 | 3433 | 3309 | 1134 |
| 8 | O | $\bar{1}$.8425 | $\bar{1}$.8182 | $\bar{1}$.7732 | $\bar{1}$.4880 | $\bar{1}$.4643 | $\bar{1}$.4572 | $\bar{1}$.4337 | $\bar{1}$.4356 | $\bar{1}$.4011 | $\bar{1}$.3887 | $\bar{1}$.1712 |
| 11 | Na | 0.0000 | 9757 | 9308 | 6456 | 6219 | 6147 | 5913 | 5931 | 5587 | 5462 | 3287 |
| 12 | Mg | 0243 | 0.0000 | 9551 | 6799 | 6462 | 6390 | 6156 | 6174 | 5830 | 5705 | 3530 |
| 13 | Al | 0.0692 | 0.0449 | 0.0000 | $\bar{1}$.7248 | $\bar{1}$.6911 | $\bar{1}$.6839 | $\bar{1}$.6605 | $\bar{1}$.6623 | $\bar{1}$.6279 | $\bar{1}$.6154 | $\bar{1}$.3979 |
| 14 | Si | 0864 | 0621 | 0172 | 7320 | 7083 | 7011 | 6777 | 6795 | 6451 | 6327 | 4152 |
| 15 | P | 1294 | 1050 | 0602 | 7750 | 7513 | 7440 | 7207 | 7224 | 6880 | 6757 | 4582 |
| 16 | S | 0.1443 | 0.1200 | 0.0751 | $\bar{1}$.7899 | $\bar{1}$.7662 | $\bar{1}$.7590 | $\bar{1}$.7356 | $\bar{1}$.7374 | $\bar{1}$.7030 | $\bar{1}$.6905 | $\bar{1}$.4730 |
| 20 | Ca | 2413 | 2170 | 1721 | 8868 | 8631 | 8560 | 8325 | 8344 | 7999 | 7875 | 5700 |
| 22 | Ti | 3187 | 2944 | 2495 | 9643 | 9405 | 9334 | 9099 | 9118 | 8773 | 8649 | 6474 |
| 23 | V | 0.3455 | 0.3212 | 0.2763 | $\bar{1}$.9911 | $\bar{1}$.9673 | $\bar{1}$.9602 | $\bar{1}$.9367 | $\bar{1}$.9386 | $\bar{1}$.9041 | $\bar{1}$.8917 | $\bar{1}$.6742 |
| 24 | Cr | 3544 | 3301 | 2852 | 0.0000 | 9763 | 9691 | 9457 | 9475 | 9131 | 9007 | 6832 |
| 25 | Mn | 3781 | 3539 | 3089 | 0237 | 0.0000 | 9929 | 9694 | 9713 | 9368 | 9244 | 7069 |
| 26 | Fe | 0.3853 | 0.3610 | 0.3161 | 0.0309 | 0.0071 | 0.0000 | $\bar{1}$.9765 | $\bar{1}$.9784 | $\bar{1}$.9440 | $\bar{1}$.9315 | $\bar{1}$.7140 |
| 27 | Co | 4087 | 3845 | 3395 | 0543 | 0306 | 0235 | 0.0000 | 0.0019 | 9674 | 9550 | 7375 |
| 28 | Ni | 4069 | 3826 | 3377 | 0525 | 0288 | 0216 | $\bar{1}$.9982 | 0000 | 9656 | 9531 | 7356 |
| 29 | Cu | 0.4413 | 0.4170 | 0.3721 | 0.0869 | 0.0632 | 0.0560 | 0.0326 | 0.0344 | 0.0000 | $\bar{1}$.9876 | $\bar{1}$.7701 |
| 30 | Zn | 4538 | 4295 | 3846 | 0994 | 0756 | 0685 | 0450 | 0469 | 0124 | 0.0000 | 7825 |
| 33 | As | 5128 | 4885 | 4436 | 1584 | 1347 | 1275 | 1041 | 1059 | 0715 | 0591 | 8416 |
| 34 | Se | 0.5357 | 0.5114 | 0.4665 | 0.1813 | 0.1576 | 0.1504 | 0.1270 | 0.1288 | 0.0944 | 0.0820 | $\bar{1}$.8645 |
| 40 | Zr | 5984 | 5741 | 5292 | 2440 | 2203 | 2131 | 1897 | 1915 | 1570 | 1447 | 9272 |
| 42 | Mo | 6203 | 5960 | 5511 | 2659 | 2422 | 2350 | 2116 | 2134 | 1790 | 1666 | 9491 |
| 45 | Rh | 0.6508 | 0.6265 | 0.5816 | 0.3024 | 0.2727 | 0.2655 | 0.2421 | 0.2439 | 0.2094 | 0.1970 | $\bar{1}$.9795 |
| 46 | Pd | 6665 | 6422 | 5973 | 3121 | 2884 | 2812 | 2578 | 2596 | 2251 | 2127 | 9952 |
| 47 | Ag | 6713 | 6470 | 6021 | 3169 | 2931 | 2860 | 2625 | 2644 | 2299 | 2175 | 0.0000 |
| 48 | Cd | 0.6891 | 0.6649 | 0.6199 | 0.3347 | 0.3110 | 0.3039 | 0.2804 | 0.2823 | 0.2478 | 0.2354 | 0.0179 |
| 50 | Sn | 7128 | 6885 | 6436 | 3584 | 3346 | 3275 | 3040 | 3059 | 2715 | 2590 | 0415 |
| 51 | Sb | 7238 | 6996 | 6546 | 3694 | 3457 | 3386 | 3151 | 3170 | 2825 | 2701 | 0526 |
| 52 | Te | 0.7438 | 0.7196 | 0.6746 | 0.3894 | 0.3657 | 0.3586 | 0.3351 | 0.3370 | 0.3028 | 0.2901 | 0.0726 |
| 56 | Ba | 7762 | 7519 | 7070 | 4218 | 3981 | 3909 | 3675 | 3693 | 3348 | 3224 | 1049 |
| 58 | Ce | 7849 | 7606 | 7157 | 4304 | 4067 | 3996 | 3761 | 3780 | 3435 | 3311 | 1136 |
| 73 | Ta | 0.8957 | 0.8714 | 0.8265 | 0.5413 | 0.5176 | 0.5104 | 0.4870 | 0.4888 | 0.4553 | 0.4420 | 0.2245 |
| 74 | W | 9029 | 8786 | 8337 | 5485 | 5248 | 5176 | 4942 | 4960 | 4616 | 4492 | 2317 |
| 77 | Ir | 9241 | 8998 | 8549 | 5697 | 5460 | 5388 | 5154 | 5172 | 4828 | 4703 | 2528 |
| 78 | Pt | 0.9289 | 0.9046 | 0.8597 | 0.5745 | 0.5507 | 0.5436 | 0.5201 | 0.5220 | 0.4875 | 0.4751 | 0.2576 |
| 79 | Au | 9332 | 9090 | 8640 | 5788 | 5551 | 5480 | 5245 | 5264 | 4919 | 4795 | 2620 |
| 80 | Hg | 9407 | 9164 | 8715 | 5863 | 5625 | 5554 | 5319 | 5338 | 4994 | 4869 | 2694 |
| 82 | Pb | 0.9547 | 0.9304 | 0.8855 | 0.6003 | 0.5766 | 0.5694 | 0.5460 | 0.5478 | 0.5134 | 0.5010 | 0.2835 |
| 83 | Bi | 9585 | 9342 | 8893 | 6041 | 5803 | 5732 | 5497 | 5516 | 5172 | 5047 | 2872 |

*A* is the element whose percentage, either weight or atomic, is being converted; *B* is the other component, usually the major one.

## TABLE 2. VALUES OF LOG ATOMIC WEIGHT RATIOS (Continued)
### Element B

| At. No. | | 48 | 50 | 51 | 74 | 77 | 78 | 79 | 80 | 82 | 83 |
|---|---|---|---|---|---|---|---|---|---|---|---|
| Symbol | | Cd | Sn | Sb | W | Ir | Pt | Au | Hg | Pb | Bi |
| At. Wt. | | 112.41 | 118.70 | 121.76 | 183.92 | 193.1 | 195.23 | 197.2 | 200.61 | 207.21 | 209.00 |
| Log At. Wt. | | 2.0508 | 2.0745 | 2.0855 | 2.2646 | 2.2858 | 2.2905 | 2.2949 | 2.3024 | 2.3164 | 2.3202 |
| At. No. | Symbol | | | | | | | | | | |
| 1 | H | $\bar{3}.9526$ | $\bar{3}.9289$ | $\bar{3}.9179$ | $\bar{3}.7388$ | $\bar{3}.7176$ | $\bar{3}.7128$ | $\bar{3}.7085$ | $\bar{3}.7010$ | $\bar{3}.6869$ | $\bar{3}.6832$ |
| 3 | Li | $\bar{2}.7906$ | $\bar{2}.7669$ | $\bar{2}.7559$ | $\bar{2}.5767$ | $\bar{2}.5556$ | $\bar{2}.5508$ | $\bar{2}.5465$ | $\bar{2}.5390$ | $\bar{2}.5249$ | $\bar{2}.5212$ |
| 4 | Be | 9044 | 8808 | 8697 | 6906 | 6694 | 6647 | 6603 | 6529 | 6388 | 6351 |
| 5 | B | $\bar{2}.9834$ | $\bar{2}.9598$ | $\bar{2}.9487$ | $\bar{2}.7696$ | $\bar{2}.7849$ | $\bar{2}.7437$ | $\bar{2}.7393$ | $\bar{2}.7319$ | $\bar{2}.7178$ | $\bar{2}.7141$ |
| 6 | C | $\bar{1}.0287$ | $\bar{1}.0050$ | 9940 | 8149 | 7937 | 7890 | 7846 | 7771 | 7631 | 7593 |
| 7 | N | 0956 | 0719 | $\bar{1}.0609$ | 8818 | 8606 | 8558 | 8515 | 8440 | 8300 | 8262 |
| 8 | O | $\bar{1}.1533$ | $\bar{1}.1297$ | $\bar{1}.1186$ | $\bar{2}.9395$ | $\bar{2}.9183$ | $\bar{2}.9136$ | $\bar{2}.9092$ | $\bar{2}.9018$ | $\bar{2}.8877$ | $\bar{2}.8840$ |
| 11 | Na | 3109 | 2872 | 2762 | $\bar{1}.0971$ | $\bar{1}.0759$ | $\bar{1}.0711$ | $\bar{1}.0668$ | $\bar{1}.0593$ | $\bar{1}.0452$ | $\bar{1}.0415$ |
| 12 | Mg | 3352 | 3115 | 3005 | 1214 | 1002 | 0955 | 0911 | 0836 | 0695 | 0658 |
| 13 | Al | $\bar{1}.3801$ | $\bar{1}.3564$ | $\bar{1}.3454$ | $\bar{1}.1663$ | $\bar{1}.1451$ | $\bar{1}.1403$ | $\bar{1}.1360$ | $\bar{1}.1285$ | $\bar{1}.1145$ | $\bar{1}.1107$ |
| 14 | Si | 3973 | 3736 | 3626 | 1835 | 1623 | 1576 | 1532 | 1457 | 1317 | 1279 |
| 15 | P | 4403 | 4166 | 4056 | 2265 | 2053 | 2006 | 1962 | 1887 | 1747 | 1709 |
| 16 | S | $\bar{1}.4552$ | $\bar{1}.4315$ | $\bar{1}.4205$ | $\bar{1}.2413$ | $\bar{1}.2202$ | $\bar{1}.2154$ | $\bar{1}.2111$ | $\bar{1}.2036$ | $\bar{1}.1895$ | $\bar{1}.1858$ |
| 20 | Ca | 5521 | 5285 | 5174 | 3383 | 3172 | 3124 | 3080 | 3006 | 2865 | 2828 |
| 22 | Ti | 6295 | 6059 | 5948 | 4157 | 3946 | 3898 | 3854 | 3780 | 3639 | 3602 |
| 23 | V | $\bar{1}.6563$ | $\bar{1}.6327$ | $\bar{1}.6216$ | $\bar{1}.4425$ | $\bar{1}.4214$ | $\bar{1}.4166$ | $\bar{1}.4122$ | $\bar{1}.4048$ | $\bar{1}.3907$ | $\bar{1}.3870$ |
| 24 | Cr | 6653 | 6416 | 6306 | 4515 | 4303 | 4256 | 4212 | 4137 | 3997 | 3959 |
| 25 | Mn | 6890 | 6654 | 6543 | 4752 | 4540 | 4493 | 4450 | 4375 | 4234 | 4197 |
| 26 | Fe | $\bar{1}.6961$ | $\bar{1}.6725$ | $\bar{1}.6614$ | $\bar{1}.4823$ | $\bar{1}.4612$ | $\bar{1}.4564$ | $\bar{1}.4520$ | $\bar{1}.4446$ | $\bar{1}.4305$ | $\bar{1}.4268$ |
| 27 | Co | 7196 | 6960 | 6849 | 5058 | 4846 | 4798 | 4755 | 4681 | 4540 | 4503 |
| 28 | Ni | 7178 | 6941 | 6831 | 5040 | 4828 | 4780 | 4337 | 4662 | 4521 | 4484 |
| 29 | Cu | $\bar{1}.7522$ | $\bar{1}.7285$ | $\bar{1}.7175$ | $\bar{1}.5384$ | $\bar{1}.5172$ | $\bar{1}.5125$ | $\bar{1}.5081$ | $\bar{1}.5006$ | $\bar{1}.4866$ | $\bar{1}.4828$ |
| 30 | Zn | 7646 | 7410 | 7299 | 5508 | 5297 | 5249 | 5205 | 5131 | 4990 | 4953 |
| 33 | As | 8237 | 8000 | 7890 | 6099 | 5887 | 5840 | 5796 | 5721 | 5581 | 5543 |
| 34 | Se | $\bar{1}.8466$ | $\bar{1}.8229$ | $\bar{1}.8119$ | $\bar{1}.6326$ | $\bar{1}.6116$ | $\bar{1}.6069$ | $\bar{1}.6025$ | $\bar{1}.5950$ | $\bar{1}.5810$ | $\bar{1}.5772$ |
| 40 | Zr | 9093 | 8856 | 8746 | 6955 | 6743 | 6696 | 6652 | 6577 | 6437 | 6399 |
| 42 | Mo | 9312 | 9075 | 8965 | 7172 | 6962 | 6915 | 6871 | 6796 | 6656 | 6618 |
| 45 | Rh | $\bar{1}.9617$ | $\bar{1}.9380$ | $\bar{1}.9270$ | $\bar{1}.7479$ | $\bar{1}.7267$ | $\bar{1}.7219$ | $\bar{1}.7176$ | $\bar{1}.7101$ | $\bar{1}.6960$ | $\bar{1}.6923$ |
| 46 | Pd | 9774 | 9537 | 9427 | 7635 | 7424 | 7376 | 7333 | 7258 | 7117 | 7080 |
| 47 | Ag | 9821 | 9585 | 9474 | 7683 | 7472 | 7424 | 7380 | 7306 | 7165 | 7128 |
| 48 | Cd | 0.0000 | $\bar{1}.9764$ | $\bar{1}.9653$ | $\bar{1}.7862$ | $\bar{1}.7650$ | $\bar{1}.7603$ | $\bar{1}.7559$ | $\bar{1}.7485$ | $\bar{1}.7344$ | $\bar{1}.7307$ |
| 50 | Sn | 0236 | 0.0000 | 9889 | 8099 | 7887 | 7839 | 7795 | 7721 | 7580 | 7543 |
| 51 | Sb | 0347 | 0111 | 0.0000 | 8209 | 7997 | 7950 | 7906 | 7832 | 7691 | 7654 |
| 52 | Te | 0.0547 | 0.0311 | 0.0200 | $\bar{1}.8412$ | $\bar{1}.8197$ | $\bar{1}.8150$ | $\bar{1}.8106$ | $\bar{1}.8032$ | $\bar{1}.7891$ | $\bar{1}.7854$ |
| 56 | Ba | 0871 | 0634 | 0524 | 8732 | 8521 | 8473 | 8430 | 8355 | 8214 | 8177 |
| 58 | Ce | 0957 | 0721 | 0610 | 8819 | 8608 | 8560 | 8516 | 8442 | 8301 | 8264 |
| 73 | Ta | 0.2066 | 0.1829 | 0.1719 | $\bar{1}.9926$ | $\bar{1}.9716$ | $\bar{1}.9669$ | $\bar{1}.9625$ | $\bar{1}.9550$ | $\bar{1}.9410$ | $\bar{1}.9372$ |
| 74 | W | 2138 | 1901 | 1791 | 0.0000 | 9788 | 9741 | 9697 | 9622 | 9482 | 9444 |
| 77 | Ir | 2350 | 2113 | 2003 | 0212 | 0.0000 | 9952 | 9909 | 9834 | 9694 | 9656 |
| 78 | Pt | 0.2397 | 0.2161 | 0.2050 | 0.0259 | 0.0048 | 0.0000 | $\bar{1}.9956$ | $\bar{1}.9882$ | $\bar{1}.9741$ | $\bar{1}.9704$ |
| 79 | Au | 2441 | 2205 | 2094 | 0303 | 0091 | 0044 | 0.0000 | 9926 | 9785 | 9748 |
| 80 | Hg | 2515 | 2279 | 2168 | 0378 | 0166 | 0118 | 0074 | 0.0000 | 9859 | 9822 |
| 82 | Pb | 0.2656 | 0.2419 | 0.2309 | 0.0518 | 0.0306 | 0.0259 | 0.0215 | 0.0140 | 0.0000 | $\bar{1}.9962$ |
| 83 | Bi | 2693 | 2457 | 2346 | 0556 | 0344 | 0296 | 0252 | 0178 | 0037 | 0.0000 |

Element A

# TEMPERATURE CONVERSION

| Cent. | Fahr. | Cent. | Fahr. | Cent. | Fahr. | Cent. | Fahr. |
|---|---|---|---|---|---|---|---|
| 0 | 32 | 175 | 347 | 350 | 662 | 525 | 977 |
| 5 | 41 | 180 | 356 | 355 | 671 | 530 | 986 |
| 10 | 50 | 185 | 365 | 360 | 680 | 535 | 995 |
| 15 | 59 | 190 | 374 | 365 | 689 | 540 | 1004 |
| 20 | 68 | 195 | 383 | 370 | 698 | 545 | 1013 |
| 25 | 77 | 200 | 392 | 375 | 707 | 550 | 1022 |
| 30 | 86 | 205 | 401 | 380 | 716 | 555 | 1031 |
| 35 | 95 | 210 | 410 | 385 | 725 | 560 | 1040 |
| 40 | 104 | 215 | 419 | 390 | 734 | 565 | 1049 |
| 45 | 113 | 220 | 428 | 395 | 743 | 570 | 1058 |
| 50 | 122 | 225 | 437 | 400 | 752 | 575 | 1067 |
| 55 | 131 | 230 | 446 | 405 | 761 | 580 | 1076 |
| 60 | 140 | 235 | 455 | 410 | 770 | 585 | 1085 |
| 65 | 149 | 240 | 464 | 415 | 779 | 590 | 1094 |
| 70 | 158 | 245 | 473 | 420 | 788 | 595 | 1103 |
| 75 | 167 | 250 | 482 | 425 | 797 | 600 | 1112 |
| 80 | 176 | 255 | 491 | 430 | 806 | 605 | 1121 |
| 85 | 185 | 260 | 500 | 435 | 815 | 610 | 1130 |
| 90 | 194 | 265 | 509 | 440 | 824 | 615 | 1139 |
| 95 | 203 | 270 | 518 | 445 | 833 | 620 | 1148 |
| 100 | 212 | 275 | 527 | 450 | 842 | 625 | 1157 |
| 105 | 221 | 280 | 536 | 455 | 851 | 630 | 1166 |
| 110 | 230 | 285 | 545 | 460 | 860 | 635 | 1175 |
| 115 | 239 | 290 | 554 | 465 | 869 | 640 | 1184 |
| 120 | 248 | 295 | 563 | 470 | 878 | 645 | 1193 |
| 125 | 257 | 300 | 572 | 475 | 887 | 650 | 1202 |
| 130 | 266 | 305 | 581 | 480 | 896 | 655 | 1211 |
| 135 | 275 | 310 | 590 | 485 | 905 | 660 | 1220 |
| 140 | 284 | 315 | 599 | 490 | 914 | 665 | 1229 |
| 145 | 293 | 320 | 608 | 495 | 923 | 670 | 1238 |
| 150 | 302 | 325 | 617 | 500 | 932 | 675 | 1247 |
| 155 | 311 | 330 | 626 | 505 | 941 | 680 | 1256 |
| 160 | 320 | 335 | 635 | 510 | 950 | 685 | 1265 |
| 165 | 329 | 340 | 644 | 515 | 959 | 690 | 1274 |
| 170 | 338 | 345 | 653 | 520 | 968 | 695 | 1283 |

| Cent. | Fahr. | Cent. | Fahr. | Cent. | Fahr. | Cent. | Fahr. |
|---|---|---|---|---|---|---|---|
| 700 | 1292 | 930 | 1706 | 1160 | 2120 | 1390 | 2534 |
| 705 | 1301 | 935 | 1715 | 1165 | 2129 | 1395 | 2543 |
| 710 | 1310 | 940 | 1724 | 1170 | 2138 | 1400 | 2552 |
| 715 | 1319 | 945 | 1733 | 1175 | 2147 | 1405 | 2561 |
| 720 | 1328 | 950 | 1742 | 1180 | 2156 | 1410 | 2570 |
| 725 | 1337 | 955 | 1751 | 1185 | 2165 | 1415 | 2579 |
| 730 | 1346 | 960 | 1760 | 1190 | 2174 | 1420 | 2588 |
| 735 | 1355 | 965 | 1769 | 1195 | 2183 | 1425 | 2597 |
| 740 | 1364 | 970 | 1778 | 1200 | 2192 | 1430 | 2606 |
| 745 | 1373 | 975 | 1787 | 1205 | 2201 | 1435 | 2615 |
| 750 | 1382 | 980 | 1796 | 1210 | 2210 | 1440 | 2624 |
| 755 | 1391 | 985 | 1805 | 1215 | 2219 | 1445 | 2633 |
| 760 | 1400 | 990 | 1814 | 1220 | 2228 | 1450 | 2642 |
| 765 | 1409 | 995 | 1823 | 1225 | 2237 | 1455 | 2651 |
| 770 | 1418 | 1000 | 1832 | 1230 | 2246 | 1460 | 2660 |
| 775 | 1427 | 1005 | 1841 | 1235 | 2255 | 1465 | 2669 |
| 780 | 1436 | 1010 | 1850 | 1240 | 2264 | 1470 | 2678 |
| 785 | 1445 | 1015 | 1859 | 1245 | 2273 | 1475 | 2687 |
| 790 | 1454 | 1020 | 1868 | 1250 | 2282 | 1480 | 2696 |
| 795 | 1463 | 1025 | 1877 | 1255 | 2291 | 1485 | 2705 |
| 800 | 1472 | 1030 | 1886 | 1260 | 2300 | 1490 | 2714 |
| 805 | 1481 | 1035 | 1895 | 1265 | 2309 | 1495 | 2723 |
| 810 | 1490 | 1040 | 1904 | 1270 | 2318 | 1500 | 2732 |
| 815 | 1499 | 1045 | 1913 | 1275 | 2327 | 1505 | 2741 |
| 820 | 1508 | 1050 | 1922 | 1280 | 2336 | 1510 | 2750 |
| 825 | 1517 | 1055 | 1931 | 1285 | 2345 | 1515 | 2759 |
| 830 | 1526 | 1060 | 1940 | 1290 | 2354 | 1520 | 2768 |
| 835 | 1535 | 1065 | 1949 | 1295 | 2363 | 1525 | 2777 |
| 840 | 1544 | 1070 | 1958 | 1300 | 2372 | 1530 | 2786 |
| 845 | 1553 | 1075 | 1967 | 1305 | 2381 | 1535 | 2795 |
| 850 | 1562 | 1080 | 1976 | 1310 | 2390 | 1540 | 2804 |
| 855 | 1571 | 1085 | 1985 | 1315 | 2399 | 1545 | 2813 |
| 860 | 1580 | 1090 | 1994 | 1320 | 2408 | 1550 | 2822 |
| 865 | 1589 | 1095 | 2003 | 1325 | 2417 | 1555 | 2831 |
| 870 | 1598 | 1100 | 2012 | 1330 | 2426 | 1560 | 2840 |
| 875 | 1607 | 1105 | 2021 | 1335 | 2435 | 1565 | 2849 |
| 880 | 1616 | 1110 | 2030 | 1340 | 2444 | 1570 | 2858 |
| 885 | 1625 | 1115 | 2039 | 1345 | 2453 | 1575 | 2867 |
| 890 | 1634 | 1120 | 2048 | 1350 | 2462 | 1580 | 2876 |
| 895 | 1643 | 1125 | 2057 | 1355 | 2471 | 1585 | 2885 |
| 900 | 1652 | 1130 | 2066 | 1360 | 2480 | 1590 | 2894 |
| 905 | 1661 | 1135 | 2075 | 1365 | 2489 | 1595 | 2903 |
| 910 | 1670 | 1140 | 2084 | 1370 | 2498 | 1600 | 2912 |
| 915 | 1679 | 1145 | 2093 | 1375 | 2507 | 1605 | 2921 |
| 920 | 1688 | 1150 | 2102 | 1380 | 2516 | 1610 | 2930 |
| 925 | 1697 | 1155 | 2111 | 1385 | 2525 | 1615 | 2939 |

APPENDIX V

# SOME THERMODYNAMIC CONSIDERATIONS

## Clausius-Clapeyron Equation

A quantitative application of the theoreum of Le Châtelier is found in the Clausius-Clapeyron equation which may be written in the form

$$\frac{dp}{dT} = \frac{Q}{T\Delta v}$$

where $dp/dT$, the change in pressure $p$ with change in temperature $T$, represents the slope of any line such as "solid-liquid," "solid-gas," or "liquid-gas" in Fig. 1-2; $Q$ is the heat absorbed or evolved per gram of substance during the corresponding phase change; $T$ is the absolute temperature; and $\Delta v$ is the change in specific volume accompanying the phase change.

This equation may be used, for example, to show that the three two-phase equilibrium curves of Fig. 1-2 must meet at point $O$ at such angles that the projection (dashed) of the solid-gas curve lies above the liquid-gas curve and the projection (dashed) of the liquid-gas curve lies above the solid-gas curve. That this is true can be deduced from the known fact that the heat of sublimation (conversion of solid to gas) is always greater than the heat of vaporization of the liquid. The volume changes $\Delta v$ in going from solid to gas and from liquid to gas at the same temperature are nearly equal, and only an infinitesimal difference in temperature $T$ need be considered if the two processes are compared close to the temperature of point $O$. Accordingly, it follows that the slopes of the two curves $Q/(T\Delta v)$ depend chiefly upon the respective heats of transformation $Q$. Thus, from the Clausius-Clapeyron equation, the slope of the solid-gas curve must be greater at point $O$ than is the slope of the liquid-gas curve, and the projection of each curve must lie above the real portion of the other curve.

## Application of the Second Law of Thermodynamics

Another, and more advanced, approach, which is found useful in the construction of phase diagrams of systems of two or more components, is the direct application of the second law of thermodynamics. It is convenient to discuss this application in terms of the free energy.[1] There is a unique value of free energy for each phase of a one-component system at each value of pressure and temperature that may be selected; in a two-component (or higher order) system the free energy varies also with the composition in such a manner that there is always a minimum in the curve of free energy

[1] A more detailed treatment of this subject is given by A. Findlay, "The Phase Rule," pp. 455–471 (revised by A. N. Campbell and N. O. Smith), Dover Publications, New York, 1951.

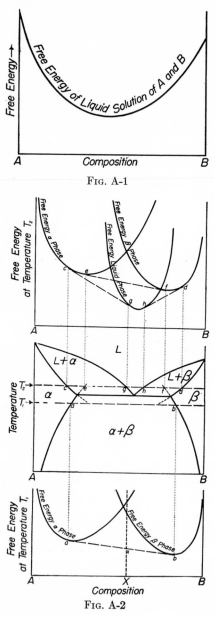

FIG. A-1

FIG. A-2

vs. composition for any fixed values of pressure and temperature (see Fig. A-1). The second law of thermodynamics, then, leads to the conclusion that *the condition for equilibrium in a system composed of one or more phases, at any stated temperature, pressure, and composition, corresponds to the condition for the minimum free energy.* In a two-component system involving two solid phases at a stated temperature $T_1$ and atmospheric pressure, there will be two free-energy curves (see the bottom diagram in Fig. A-2). Since free energies are additive, the minimum free energy for an intermediate composition will lie on the straight line drawn tangent to the two free-energy curves. The lowest free energy for each alloy of the system $AB$ will then lie on the $\alpha$ curve from pure $A$ to composition $a$, on the tangent line from $a$ to $b$, and on the $\beta$ curve from $b$ to $B$. An alloy of composition $X$ will have the value of free energy indicated at $x$, which lies below either the $\alpha$ or $\beta$ free-energy curves and therefore corresponds to the most stable condition of alloy $X$, namely, a mixture of $\alpha$ and $\beta$ in proportions indicated by the lever principle. The points $a$ and $b$ designate the compositions of the $\alpha$ and $\beta$ phases, respectively, which can coexist at equilibrium at temperature $T_1$ in the binary eutectic diagram of Fig. A-2.

Consider next the free-energy curves corresponding to a higher temperature $T_2$, top diagram in Fig. A-2. Here a third free-energy curve corresponding to the liquid phase has been introduced. It extends below either of the solid-phase curves, because liquid is stable at intermediate compositions at this temperature. Equilibrium between the $\alpha$ and $\beta$ phases is seen no longer to be possible, because all of the line $ef$, tangent to the $\alpha$ and $\beta$ curves, respectively, lies above the line $cghd$. In its stead there is equilibrium between the $\alpha$ and the liquid phases from $c$ to $g$, stability of liquid from $g$ to $h$, and equilibrium between the liquid and $\beta$ phases from $h$ to $d$. It will be evident from the diagram that for geometric reasons point $c$ must always lie to the left of point $e$ and that point $d$ must always lie to the right of the point $f$ as long as the free-energy curve of the liquid projects below the line $ef$. Therefore, the boundaries of the $\alpha + \beta$ field in the phase diagram (center of Fig. A-2) must always be drawn so that they project into the $L + \alpha$ and $L + \beta$ fields, respectively.

# INDEX